001.422 H782h

Hooke, Robert, 1918-

How to tell the liars from
 the statisticians

001.422 H782h

Hooke, Robert, 1918-

How to tell the liars from
 the statisticians

How to Tell
the Liars from
the Statisticians

How to Tell the Liars from the Statisticians

ROBERT HOOKE

Statistics Consultant
Pinehurst, North Carolina
Formerly, Manager, Mathematics
Westinghouse Research and Development Center

with illustrations by JOSEPH M. LILES
Professor of Art
North Carolina School of Science and Mathematics
Durham, North Carolina

MARCEL DEKKER, INC.   New York and Basel

Library of Congress Cataloging in Publication Data

Hooke, Robert [date]
  How to tell the liars from the statisticians.

  (Popular statistics; 1)
  Includes index.
  1. Statistics.   I. Title.   II. Series.
QA276.12.H66 1983   001.4'22   82-23593
ISBN 0-8247-1817-8

Marcel Dekker, Inc.
270 Madison Avenue, New York, New York 10016

Current printing (last digit):
10 9 8 7 6 5 4

Printed in the United States of America

For
Annis
who helped and insisted

. . . it is clear that nobody who does not understand insurance and comprehend to some degree its enormous possibilities is qualified to meddle in national business. And nobody can get that far without at least an acquaintance with the mathematics of probability, not to the extent of making its calculations and filling examination papers with typical equations, but enough to know when they can be trusted, and when they are cooked.

George Bernard Shaw
"The Vice of Gambling
and the Virtue of Insurance"

Preface

Liars, as portrayed on screen and in fiction, are often charming rogues, while statisticians are always persons of infinite dullness. In real life, this is not the way you tell one from the other.

Faces sometimes chosen for ability to project sincerity flash onto our television screens, toss a few facts and numbers at us, and quickly vanish. They seem to be confident that we'll agree that their numbers can lead to only one conclusion and most viewers seem to believe them. The few doubters include those with inside knowledge, congenital skeptics, and statisticians, who are left saying, "Yes, but . . . ," as the next commercial appears.

The science of statistics has made great progress in this century, but progress has been accompanied by a corresponding increase in the misuse of statistics. The public, whether it gets its information from television, newspapers, or newsmagazines, is not well prepared to defend itself against those who would manipulate it with statistical arguments. Many people either believe everything they hear or come to believe in nothing statistical, which is even worse.

Good statistical practice is an absolute necessity to any advanced society, and we can't afford to neglect this valuable tool just because some people misuse it. This book was written in the belief that informed citizens, with or without any interest in statistics per se, or in mathematics or proba-

vii

bility in general, can enjoy learning some of the ways of distinguishing good statistical reasoning from bad.

This book, simply and without formulas, shows how statistical reasoning affects nearly all aspects of our lives. It touches on drug testing, discrimination, sports, political polls, compulsive gambling, gun detectors, cancer research, crime and punishment, opinion surveys, advertising, mass production, and doctors' waiting rooms. These and a host of other examples are used to show that statistics, far from being a dull subject about collections of numbers, is one of great interest to almost anyone who ever has hard decisions to make.

Robert Hooke

Contents

Contents

Contents

Contents

Contents

POPULAR STATISTICS

a series edited by

D. B. Owen
Department of Statistics
Southern Methodist University
Dallas, Texas

Nancy R. Mann
Biomathematics Department
UCLA
Los Angeles, California

1. How to Tell the Liars from the Statisticians,   *Robert Hooke*

In preparation

2. Educated Guessing: How to Cope in an Uncertain World, *Samuel Kotz and Donna Fox Stroup*

About the Series

Statistics is the essence of the scientific method – it is used across all major disciplines. The aim of the Popular Statistics series is to present individual and unique volumes written with the nonmathematically oriented reader in mind. Although statistically oriented toward applied areas such as business, technology, science, medicine, law, and economics, these volumes require no prior statistical training and are "ready to read."

It is hoped that these books will be read and used in the same spirit that they were written – with great enthusiasm and a desire to increase knowledge of the power of statistics.

D. B. Owen
Nancy R. Mann

1

Statistics Are — Statistics Is

Almost everyone has heard that "figures don't lie, but liars can figure." We need statistics, but liars give them a bad name, so to be able to tell the liars from the statisticians is crucial.

It is commonly believed that anyone who tabulates numbers is a statistician. This is like believing that anyone who owns a scalpel is a surgeon. A statistician is one who has learned how to get valid evidence from statistics and how (usually) to avoid being misled by irrelevant facts. It's too bad that we apply the same name to this kind of person that we use for those who only tabulate. It's as if we had the same name for barbers and brain surgeons because they both work on the head.

Most people think of statistics as plural: collections of numbers and little else. To statisticians, statistics is singular: a fascinating subject that relates to almost everything we do. A quick glance at the index will support this position. Statistics-singular deals with things other subjects dismiss as unpredictables ("your mileage may vary"), with chance and choice and trade-offs, with the basis of government policy, with cause and effect, and so with the very essence of science.

The formal educational process provides very little information about statistics-singular to most students. This leaves the students vulnerable to those individuals I call the data pushers, who, somewhat like dope pushers, try to gain control over us with their product. "Liars" may be too

1

strong a word for them, since it suggests falsification of data. The data pushers include, in descending order of maliciousness, those who deliberately try to deceive us with true but misleading data, those whose enthusiasm for a cause leads them to do this unconsciously, and those who merely combine their misinformation with persuasiveness.

A kind of perverse enjoyment can be had from watching the data pushers and spotting the flaws in their arguments. This spectator sport, like all others, requires a little knowledge of what the players are doing, and that's what I've tried to provide. All of us would be better off if more of us would acquire the habit of reading or listening critically when people are quoting numbers.

History is supposed to teach us how to deal with the present and future, but it doesn't do that if we look at it merely as a record of events. So also statistics-plural don't teach us much if we look at them as dead records of the past—when you've seen ten numbers you've seen them all. But statistics-singular has to do with how we use numerical records to deal with the chanciness of our lives. This is my subject, together with some digressions into related areas where misconceptions based on numbers abound.

2

2
Worry

An individual who is afraid to go outdoors for fear of being hit by a meteorite is generally regarded as neurotic. Such worry is unreasonable, we think, because being hit by a meteorite, while quite possible, is an exceedingly rare event. Being struck by lightning is also rare, but it is more likely than being hit by an astronomical object, so we devote a little more concern to it. Being struck by a car is still more likely, so we are careful when we cross streets, but the chances are still low enough that we don't give much thought to the possibility of being run down.

The variety of disasters that can occur to us is so enormous that we tend to preserve our sanity by not thinking about most of them. We budget our worry, or concern, over the most probable ones and let the others go. Even the most rational people, however, fail to do this in the best way and usually have some areas in which their emotions take over and distort the probabilities. This is where the data pushers come in. One of their strongest weapons consists of playing on our worries, especially the irrational ones.

The prime goal of the data pusher who wants to sell us some protection against being hit by a meteorite is to make us forget that such a possibility is considerably more remote than catching a cold. We can resist the data pusher's wiles either by learning more about meteorites or more about recognizing misleading data. Since meteorites are just one of millions of possible subjects, the latter course is easier.

3

The data pusher's job is made easier by the fact that there are so many people who shy away from numbers as if they were poisonous reptiles. Probabilities are expressed equivalently as decimals, fractions, or percentages, and fear of these things makes an individual an easy mark for the persuaders. In this book, for the purpose of citing examples, a probability or a chance will occasionally be expressed numerically, but otherwise every effort will be made to avoid offending those who are repelled by numbers.

3
More on Worry

There is a dread disease called bilharziasis. (Australian soldiers in World War I called it "the Bill Harris" rather than struggle with the pronunciation.) This very serious disease is caused by a little worm called a blood fluke, and if you read about the horrible symptoms you may want to dash off and pay $20 for an inoculation, if there is such a thing. On the other hand, paying $20 in an attempt to ward off every disaster we hear mentioned is to court the certain disaster of poverty.

When the butterflies in my stomach tell me it's time to worry about some particular possibility, I try to look at two aspects of it: how likely is it to happen, and what are the consequences if it does happen? Of course, the worse the consequences the more I'll worry, even if the likelihood is very small. A favorite device of data pushers is to put so much emphasis on the consequences that you may forget that the worrisome event virtually never happens.

How likely are you to be afflicted with bilharziasis? Actually, if you avoid walking in African streams, you won't catch this disease, but if I were selling inoculations I might not tell you that.

Sensible people take precautions against possible disasters, and of course one man's idea of precaution is another man's silly worry, but whether we're heavy worriers or light worriers, we all have to draw the line somewhere, and say, "This thing happens too rarely to give me concern."

4
Statistics and Chances

"More accidents happen around the home than anywhere else." This is a statistic. It doesn't quote any numbers, but numbers can probably be unearthed to support it. With or without numbers, it is not a very informative statistic. The only conclusion it can possibly lead to is that home is the most dangerous place to be, which is obviously false. "Most car accidents happen within x miles of home," we're often told, x being some small number. Does that mean that we would be wiser to garage the car x miles from home and walk back and forth to it? Undoubtedly not, except for the possible benefits of the exercise and saving of gasoline.

A little-known fact about the subject of statistics is that it has mostly to do with probability, or chances. The accident statistics just quoted are not useful because they don't tell us anything about our chances. People have more accidents at home than anywhere else because home is where many of them are. More car accidents occur within x miles of home because most of the driving occurs there. If you really care about your chances of having an accident within x miles of home, you have to find statistics that take into account the amount of time spent or the number of miles driven within x miles of home. Such refinements may be very hard to obtain, but don't give anyone part credit for quoting useless, unrefined statistics just because meaningful statistics are hard to obtain. If a space program fails to put a man on the moon, we give no part credit for putting him up into a tree.

A television news show reported a campaign to raise the minimum age for school bus drivers in a state where teenage drivers are allowed. Statistics were presented that showed that 19-year-old drivers had had more accidents than 20-year-old drivers. Nothing was said about how many drivers there were of each age, so the statistic was useless for drawing conclusions, but most viewers probably did not notice that. It may well be that by age 20 most of the better drivers have gone off to greener pastures. In fact the original reason for permitting teenage drivers was that it was hard to find responsible older people who would do this futureless and often unpleasant job.

Statistics are history, and we gather them so that we can learn from history. Usually what we can learn is how our chances of success depend on our actions. If a statistic doesn't give us information on these chances, it is likely to be useless, or worse.

5
Measuring Chances

The weather bureau used to say, "It will probably rain today." Some years ago they began to give more information, and now they say, "The probability of precipitation today is 60%."

This change divided the population into two groups: those who saw the change as an improvement, and those who complained about it and even claimed that the new statement was less informative than the old. The problem with most of the latter group is that they dislike, or even fear, numbers. Recently such people have been dubbed "innumerates," meaning that their relation to numbers is similar to that of illiterates to words. It is a habit of some innumerates to try to cover up their inadequacy by proclaiming it to be a virtue, but we shall do as the weather bureau did and plow ahead, hoping in time to wipe out innumeracy.

Just as we use numbers to say how much more this person weighs than that person, or how much hotter this day is than that day, so we also use numbers to say how much more likely this outcome is than that outcome. Whether we speak of chances, probability, or likelihood, we are talking about events whose outcomes are uncertain. All of us are constantly affected by the uncertainty of events, and we make decisions based on what we think the chances of various outcomes are. Numbers help us to do this.

If a baseball player has a batting average of .300, this is a statistic and

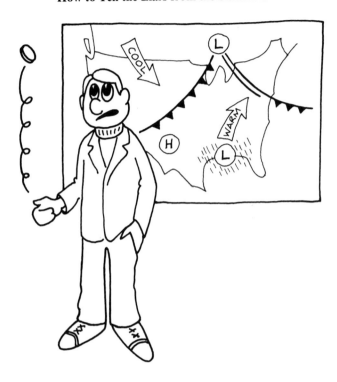

an historical fact. It also means that the player's chance of getting a hit when he comes up is about 30%, and this is how the statistic is used. The manager, in selecting a pinch hitter, will mentally compare the averages of the hitters. (He will probably also take into account other factors, such as whether the pitcher is left-handed or right-handed, but the batting average is a starting point in the decision process.)

Most card players and sports fans are well acquainted with numerical ways of measuring chances, so it is sometimes easier to explain a point of probability to them than to a brain surgeon or mechanical engineer who has never wasted time on such frivolities. However, not all card players or sports fans can transfer ideas about measuring chances from games to more serious matters. In the fifth grade I had friends who could not understand decimals at all, even though they could all compute their batting averages. They simply weren't motivated by abstract decimals. I think the popularity of the football phrase "on any given Sunday" reflects the unwillingness of many fans to deal with numerical odds. Early in the history of the National Football League, when virtually every game was played on Sunday, it became popular to note that on any given Sunday even the worst of the teams might come up with a victory. This became such a cliché in media postmortems that it is now used only in its abbreviated form "on any given Sunday" and every football fan mentally fills in the rest. The cliché is true, but its information content is low; it seems to be popular with people whose view of life can be summed up simply in the phrase "anything can happen."

So little probability is taught in our schools that most people know practically nothing about it, and their ignorance leads to ineptness in understanding statistics. The probability that is taught is usually about very simple things such as cards or dice. If you draw five cards from a well-shuffled deck, a mathematician (or an almanac, for that matter) can tell you that the odds against drawing a straight are 254 to 1. If you step outside your front door, no one can tell you the exact odds against your being struck by a meteorite. Some would like to conclude from this that probability isn't of much use when it departs from card games and looks at real life, but this isn't true — if you can't distinguish between your chance of being killed by a meteorite and your chance of dying of lung cancer after smoking two packs of cigarettes a day for many years, then you are a prime target for the data pushers.

6

Chances and Drug Laws

"Use of this drug may in some individuals cause headaches, nausea, hypertension, or, in extreme cases, death."

When such warnings first began to appear, it seemed that the pharmaceutical professions were trying to provide us with useful information. However, it soon developed that they had almost identical warnings to accompany virtually every drug, so that it became obvious that these statements were mere disclaimers of legal responsibility.

I say this because a warning of possible side effects is useless unless it says something about the odds against experiencing them. When smallpox vaccinations were first given, some people died of the inoculations. Nevertheless, people gladly took the vaccinations because their survival chances were better with them than without them. But as smallpox became less and less prevalent, the death rate from the inoculations had to be reduced in order to keep the cure from becoming worse than the disease.

If a side effect is temporary, like a headache, it may be true that the only warning needed by the patient is that any subsequent headache may be due to the drug. But if the side effect is permanent, like death, the patient needs to know the odds: does this drug result in the death of one user in a million or one in ten? Without knowing the odds, the patient and the doctor cannot sensibly decide whether or not the risk is worth taking.

7
The Broad-Base Fallacy

Possibly I have already told you more about bilharziasis than you wanted to know, but it's a useful example — I can pretend to be selling anti-BHZ inoculations, and there is no BHZ foundation around to take exception to what I say. Suppose I find that 1 in 10,000 Americans go to Africa annually and die from the disease. This figure would not be helpful in my promotion because most people have enough to worry about without concerning themselves with 9999-to-1 shots.

What I would do, as an imaginary purveyor of fear of BHZ, would be to multiply the 1 in 10,000 figure by the population of the United States, which I will take to be 200 million in round numbers. I could then report back to you that 20,000 Americans die annually of BHZ. The thought of the accumulated misery of 20,000 victims and their families produces an entirely different feeling from that generated by the 9999-to-1 odds. Neither figure is always the *right* one to quote. The odds should be considered in deciding whether or not to have an inoculation, especially if the inoculation has any undesirable side effect; the total is the proper statistic to be considered if we are asked to give money to a foundation for the relief of human misery.

The point here is that in a country of 200 million people any very rare event can be made to sound commonplace by the simple act of telling how many people it happens to. I call this "the broad-base fallacy" because data pushers sometimes use it to appeal to our emotions in cases where we should be considering the odds.

A real example of this fallacy is in traffic fatality statistics. The media emphasize the total number of deaths and have thereby managed to convey to the American public the impression that travel on U.S. highways has become steadily more dangerous over the years. Actually the opposite is true, as can be seen from the continuously decreasing death rate per vehicle or per passenger mile. Overemphasizing totals can lead us to make false trade-offs. If traffic continues to increase, misplaced emphasis on total fatality figures can lead us to legislate lower and lower speed limits; this can cause slower deliveries, higher prices for food and other items, greater traffic jams and pollution, and delays for people trying to reach hospitals. It's a common notion that price increases are not to be considered in discussing fatalities, but this notion is popular only among the affluent, for whom price increases are never fatal.

While on the subject of traffic statistics, this is an appropriate time to mention another media favorite, the one that enables them to write colorfully of the "holiday carnage" on the highways. This is the figure that tells how many will be or were killed in traffic accidents on a given holiday weekend. In the first five years of the national speed limit of 55 miles per hour, the greatest number of people killed during any four-day holiday was 713. In a year of 48,000 fatalities, the average number for *any* four-day period is 526. Since 713 is only 36% larger than 526, then if that worst four-day weekend involved a traffic increase of at least 36%, the highways were no more dangerous than on average days.

8
The Unmentioned Base

A company that makes only 1 or 2% profit on sales may be in a precarious position, because such a profit can be quickly wiped out by a slight increase in costs. But if the company made 1% profit last year and increased it to 2% this year, headlines will scream about "a 100% increase in profits," leading the casual reader to form an impression of gross excess profit taking. The headline has shifted the base of the percentage from sales to profits, and such base shifting is an easy way to create false images. By the same argument, a company that goes from a 0% profit to a 2% profit could be accused of making an infinite increase, but this is such an obvious exaggeration that no one tries it.

Every percentage is a percentage *of* something, and this something is called the base. Since the base is often understood and not mentioned, changing it to cause confusion is easy to do. An incumbent president, for example, can be accused of causing an average rate of inflation of 15% a year; but if the rate was 14% this year as opposed to 16% last year, the incumbent will boast of having reduced inflation by 12½%, which is what a drop of 2 percentage points out of 16 works out to. If a lot of voters believe that a 16% inflation rate reduced by 12½% is down to 3½%, the incumbent will not be displeased. Such voters fail to ask the always-crucial question, "percent of what?"

Other ways of changing bases abound. An easy one is to abandon the

percentage, when it is the appropriate figure, and go to totals or to a rate per some small unit. Totals are used when it is desired to make a change look large, as in the broad-base fallacy, while rates per some small unit are used to minimize a change. In a period of 10% inflation, a union may ask for a 15% raise, but if the average hourly wage is $6 the union will describe it as a 90¢ raise, this being the average raise per hour. Reducing anything to less than a dollar always creates an image of smallness. The company, on the other hand, will offer a 10% raise and will point out that for its 100,000 employees working 2000 hours a year this comes to an increased cost of $120 million. The bystander who hasn't kept his eyes glued to the base may find it hard to follow the reasoning that makes $120 million less than 90¢.

For another example, let's take a utility company that wants to keep up with 10% inflation by raising its rates 10%. Suppose it has a million customers with an average monthly bill of $72. A business reporter can choose from these three headlines:

Utility Asks 1¢ Rate Increase
Utility Seeks 10% Rate Hike
Utility Asks for $86.4 Million Increase

The first of these is based on the hourly rate, as is done when unions demand raises. The second is the reasonable way of describing the request. The third is an annual total and is calculated to encourage angry readership. It isn't necessary to point out which one of these three headlines is nearly always the chosen one.

9

The Elmer Gantry Effect

In 1927 Sinclair Lewis wrote a popular novel about Elmer Gantry, a hypocritical preacher. This book probably opened the eyes of some readers to the possibility of such a person, and it may have helped some readers to recognize charlatans. But the book also helped to spread the word that rascals tend to make more interesting fiction than saints. Since then, fictional preachers who are bent on leading pretty choir singers astray probably outnumber their more well-behaved colleagues by 3 to 1.

Fiction provides a way for false statistics to enter our lives so stealthily that we scarcely notice them. Reading fiction, of course, provides us with insights into human nature from the life experiences of hundreds of authors, and to this extent fiction helps the reader to build up an experience with the thinking of other people that could not be acquired by a single individual's lifetime contacts with real people. But there is a deep-seated bias here. Because the activities of scoundrels make good stories, they are even more common in fiction than in real life. People who fail to notice this fact tend to conclude that preachers are mostly hypocrites, businesspeople are mostly crooks, scientists are mostly mad, and private eyes are all smarter than policemen.

Racial and other stereotypes in fiction occur from the same effect. An Italian should not be offended if a fictional Italian belongs to the Mafia. Perhaps he or she should not even complain if a fictional Mafia group consists mostly of Italians. But he or she does have a legitimate gripe if the majority of Italians in fiction belong to the Mafia, because this creates an impression that is demonstrably false.

Unfortunately there's no way to control the frequency of occurrence of various character types in fiction. If Sinclair Lewis can write about a hypocritical preacher, any other author has the right to do likewise. As readers we have to exercise our reason and remember that, though a novel can expose the possibility of some kind of human behavior, a group of novels can't tell us how often that behavior occurs in real life.

10

When the Truth Is not
the Whole Truth

If a manufacturer says, "90% of the cars we've ever made are still on the road," this may be exactly true, but its meaning is up for grabs until some other facts come to light. The whole truth includes telling how long the company has been making cars — if it has been in business for only the last four years, the claim lacks significance. Even if it has been going for thirty years, it may be a growing company, so that most of the cars it has built are less than four years old.

A 1947 survey of Chicago lawyers showed that those with college degrees had an average annual income of $11,373, while those who were only high school graduates earned $12,095.* Did this mean that education diminishes the value of the lawyer? That educated lawyers were discriminated against? That clients preferred lawyers in the same educational group as themselves? None of the above.

The whole truth is that the college graduate lawyers were younger than the others. The practice of obtaining a college degree before going into law was much less prevalent in the years 1900–1930 than it became later. As a result, in the year 1947 college-educated lawyers tended to be younger, less experienced, less well known, and so less well paid. A breakdown of the two groups of lawyers by year of admission to the bar showed that in

*W. A. Wallis and H. V. Roberts, *Statistics: A New Approach* (Glencoe, Ill.: The Free Press, 1956), p. 298.

almost every age group the college graduates made more money than the others.

Once we know the whole truth, no one can sell us on the idea that a college education is a detriment to a lawyer's career. But if there were a special interest group dedicated to this end, they would keep bombarding us with various forms of this same fallacy, always pointing out that the college education accompanied the diminished income, and never mentioning the crucial role played by age and experience.

11

Ceteris Paribus

Ceteris paribus means "other things being equal," and I drag in this bit of
Latin to suggest the antiquity of the phrase. Often we assume that other
things are equal without saying so, and then we may forget that the assumption was made. Data pushers like to profit from our habit of making
this tacit assumption.

During the 1970s a report was issued describing data for crude cancer
mortality rates in two groups of ten American cities, one group that had
added fluoride to the drinking water and one that had not. Between 1950
and 1970 the cancer mortality rate rose in both groups of cities, but it rose
considerably faster in the fluoridated group. Some data pushers were more
than willing to attribute the difference to the fluoride, without pointing
out that they were assuming "other things being equal."

The National Cancer Institute asserted that indeed other things were
not equal and that the difference between the two groups of cities was explained by differences in the age, race, and sex distributions of the two
groups of cities; these three demographic characteristics all have known
effects on cancer mortality rates. In Great Britain the Royal College of
Physicians asked for comment from the Royal Statistical Society. Two
statisticians were chosen to analyze the data; here is what they found.

The cities that were to be fluoridated already had, before fluoridation, an excess of cancer deaths over the "control," cities that went un-
fluoridated. Between 1950 and 1970 the demographic differences between

21

the two groups became even greater. Taking into account the known effects of age, race, and sex on cancer death rate, they found that the remaining unaccounted-for part of the cancer mortality rate was actually somewhat smaller in the fluoridated cities than in the others.*

Which all goes to show that *ceteris* are not necessarily *paribus,* and that the reason things are often not what they seem is that we may be assuming things about them that are not true. It's worth noting that in the case of the lawyers' earnings, the issue is noncontroversial, and the additional breakdown quickly settles the question. In controversial issues, however, we can expect the data pushers to keep pressing their case. Undismayed by being proved wrong, they often continue fighting by presenting their half-truths to new audiences who haven't been exposed to the less spectacular parts of the truth.

---

*P. D. Oldham and D. J. Newell, "Fluoridation of Water Supplies and Cancer — A Possible Association?," *Applied Statistics* (*Journal of the Royal Statistical Society, Series C*), vol. 26, no. 2 (1977).

12
Statistics of Discrimination

I embark on the subject of discrimination with considerable trepidation, since it is so charged with emotion that the most innocent and objective observations lead to accusations of prejudice and bigotry. For the same reason, there is no area in which statistics are more widely misused and misinterpreted. I happen to believe that a good cause deserves good statistics and am put off by being lied to or by having my intelligence insulted, even for a good cause.

A flagrant example is one which is widely used, at this writing, to interest people in the problems of working women. Government figures released in 1980 on earnings of full-time workers showed that the median income for women was only 59% of the median for men. Since then the number 59% has been shouted from the rooftops with the implication that it is entirely due to discrimination by employers. Some of it obviously is due to discrimination and that is something that most of us would like to see eliminated. But there are many other factors involved that employers can't change. Here are a few:

1.   In recent years women have been entering job markets in rapidly increasing numbers. It follows that in 1980 working women had, on the average, much less experience than men. Any earnings figures that fail to provide a breakdown by age and experience should not be considered meaningful.

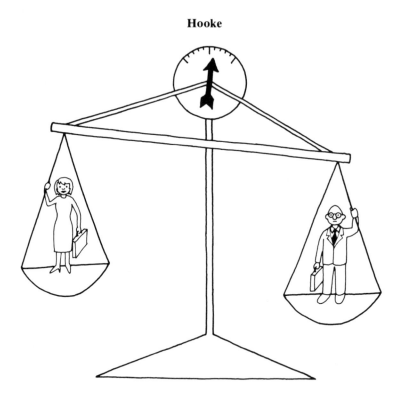

2.   It has been pointed out elsewhere that women have tended to choose jobs such as teaching and nursing, which seldom lead to very high salaries. They flock into computer programming and avoid the much more lucrative field of computer sales. In the past these choices have sometimes been forced on them, but unless and until women can diversify in the same way as men, overall statistics will remain meaningless.

3.   Until recently it was traditional for the husband to choose where he wanted to work and for the working wife to find the best job she could nearby. He usually made himself more valuable by being willing to travel, while she was often reluctant to leave home, for various reasons. When he was promoted and relocated, she often started from scratch in a new city. The tradition is dying out, but its depressing effect on average women's salaries will be around for a while and is not something that employers or legislators can change.

4.   Since we are talking about averages, it should be pointed out that, while many women are as wrapped up in their careers as any man, the averages have been affected by many other women who worked toward

more temporary goals. The latter naturally drag down the earnings median for women, and this will continue to be true until the percentage of men with this kind of motivation becomes the same as that of women.

It isn't easy to find good salary data that take into account important variables such as age, experience, educational qualifications, and job content, but it's clear that with such variables in the picture it's illogical to blame everything on sex.

13

A Paradox in Discrimination
Statistics

A well-known university was accused of discrimination against women who applied for admission to its graduate school. Simple statistics showed that the acceptance rate for women applicants was definitely lower than that for men. Since prospective graduate students apply to a particular department, and since each department selects its own applicants to be admitted, it appeared to be a simple matter to rectify this situation by finding out which departments were guilty of discrimination. But a breakdown of the statistics by department showed, paradoxically, that no department was discriminating.

Most people's immediate reaction to such a finding is to deny the possibility that every department can be fair while their total appears to be unfair. To show that it is possible, all that we have to do is produce one example in which it happens. Rather than go into the extensive data of the real case, I will give here a simple hypothetical example that is easy to follow:

Graduate School Admission by Sex, Hypothetical School

	No. applied	No. admitted	% admitted
Department A			
Women	50	25	50%
Men	100	50	50%
Department B			
Women	100	30	30%
Men	50	15	30%
Total school			
Women	150	55	37%
Men	150	65	43%

Here we see that 37% of women applicants to the whole graduate school are admitted while men are admitted at a 43% rate. Yet each department admits men and women at the same rate.

The explanation is that men and women do not apply to various departments at the same rate. In a real school some departments are much harder to get into than others, just as in this hypothetical example. Note that department A admits 50% of its applicants while department B can find room for only 30% of those who apply. If it turns out that women apply in greater numbers to those departments that are harder to get into, for example English as opposed to Engineering, then even if every department is scrupulous about admitting women at the same rate as men, the overall total does not reflect this fact.

This example may seem to violate the axiom that "the whole is equal to the sum of its parts," but it really does not. The admission rates of men and women can be the same in each part but not over the whole because an overall rate is not the sum or average of component rates. Forgetting this simple fact about rates is a common reason for misinterpretation of data.

14
Smoking and Cancer

When statistical links between smoking and cancer first began to appear in the 1950s, two opposing groups of data pushers immediately formed. When it was seen that smokers had a great deal more lung cancer than non-smokers had, one group of analysts wanted to jump immediately to the conclusion that smoking caused the cancer. The opposing group pointed out that smokers and nonsmokers are not otherwise equal—nonsmokers have different personalities to begin with, demonstrably so because they are able to withstand the blandishments of tobacco companies. Their environments are also not otherwise equal—people under stress have a tendency to become smokers, and smokers are more prevalent in cities, where pollution confuses the issue. All these factors, and others, could possibly account for the extra amount of cancer seen in the smoking population.

The former group turned out to be right, but only because bad logic does not necessarily lead to wrong conclusions. The latter group took the proper position, if we exclude those members of the group who acted as if they had proved the innocence of smoking tobacco after they had shown the existence of other possible culprits. Eventually the correct steps were taken: after the statistics suggested that smoking could cause cancer, scientists went to work and found the underlying causes. The combination of good statistics and good science brings the nonbelieving group down to the irreducible minimum. It is everyone's privilege to believe as he or she wishes, and that's why we have a Flat-Earth Society and a group who believe that we never put a man on the moon.

15

People Are Different

Two lengths of copper wire, of the same diameter, will differ only very slightly in their ability to carry electric current. Two human beings of the same age, sex, race, color, height, weight, and religious belief may differ enormously in their ability to tolerate alcohol.

One thing that all of us (most of us, anyhow) learn at an early age is that people are different. Yet so-called experts in various fields continue to make statements that are alleged to be true of all people, and many of their statements are widely believed, sometimes in spite of much evidence to the contrary.

The great success of the experimental method in the natural sciences has led to its increased use in other areas, and this is good, up to a point. If a certain combination of conditions always leads to the same result, it is apt to be fairly easy to set up those conditions and the nature of the result can be determined by observation. In the natural sciences there are few, if any, such simple cases left that haven't been fully explored. In medicine, psychology, sociology, and other areas that deal with people, there never were any such simple situations.

So beware of any simple experiment involving people that comes up with a simple conclusion. If you are told that all overweight is a result of overeating, don't believe it. (Fifty years ago most people didn't believe it. Then we went through a period about twenty-five years ago when we were

told to believe that scientists had established the truth of this statement. Now we hear about a lot of exceptions again.)

If we had a headache remedy that cured every headache very quickly, it wouldn't take much of an experiment to convince us of its powers. What we have, though, are headache remedies that cure some headaches in some people some of the time, alleviate some, and provide no help at all for others. Replace the word "headache" with anything from indigestion to cancer and you will still have a true statement. Also, as pointed out before, life is complicated by the fact that some patients recover without any treatment. Side effects, if anything, are of an even more complex nature.

What does all of this have to do with data pushing? Seizing on the common desire for simple solutions, data pushers often perform simple experiments and claim simple positive results that may lead us down the garden path to confusion. Everybody should know a little about the business of experimentation; more about this later.

16
Risk Taking in Industry

It has been said that during the lifetime of a corporation division its leader should be in turn a risk taker, a caretaker, and an undertaker. The first develops a business where none existed before, the second nurtures it over the years of its success, and the third keeps it as profitable as possible while it dwindles into obsolescence. Most personalities are not adaptable enough to do each at the right time, so these roles are usually played by at least three separate individuals.

Large corporations sometimes have trouble finding the risk takers, because a corporation, in speaking of risks, means risks to the corporation. Boards and executives may fail to take note of the fact that while managers also *speak* of risks to the corporation they *think* in terms of risks to their own job security. The corporation often needs to enter a number of fields in each of which the probability of success may be low, but where successes in one or two fields more than compensate for the overall losses. The individual manager will see the sense of this, but will note that most of the managers who fail are fired or demoted. This observation encourages the manager to look for the small but relatively sure profit at the expense of the less likely large one that the corporation wants.

Corporate line managers also tend to become conservative because the "caretaker" part of a business's history is usually the longest part. During that period profits come in so long as no one "rocks the boat," and when profits are made, even if they are not record-breaking, the boss tends to keep his or her job.

Unfortunately, conservatism tends to grow not only in the routine production parts of the corporation, but also in research and advertising. In these two fields new ideas are important. A person who has one good idea and ten bad ones may be a success here, as opposed to the line manager who can have ten good ideas overwhelmed by one bad one. But as research and advertising become big enough to require their own line management, conservatism sets in. In case anything goes wrong, line managers like to be able to say that what they did was normal and orthodox, even though in some areas the normal and orthodox bring on the undertaking phase prematurely.

Managers are placed into their positions because they are supposed to be, at least by comparison to others, wise and capable of making good decisions. Job insecurity, however, leads them to look for every conceivable way of avoiding decisions. History abounds with stories of various scapegoats that have been used by decision makers when they are attacked by second-guessers. A common one is tradition, the scapegoat implied in the excuse, "I went by the book." A more contemporary refuge is, "I only did what the computer said to do."

17
The Risk Taker's Image

If I put my money into a gold mine that quickly runs out of gold, I will probably be labeled "a speculator who frittered away his fortune." If the mine produces even more gold than was expected, I become "an aggressive investor who built his small fortune into a great one." The outcome may be entirely independent of my astuteness as an investor, because in spite of all scientific methods of estimating, the amount of gold in an uninvestigated spot remains a mystery.

People do tend to apply hindsight in judging the strategies of others, and while this may sometimes be unfair it is sometimes more just than it might appear. The same kind of investment that is frittering away a fortune for one person can be an aggressive investment for another. The difference lies in the amount of capital that each individual has available.

Suppose it takes a million dollars to explore and develop a mine. Suppose further that only about one in twenty such explorations lead to any substantial amount of gold, but that when there is gold it is worth a hundred million or more. Then an investor with only one million dollars to invest will almost surely lose it all, since the odds are 19 to 1 against finding anything. On the other hand, an investor with many millions can keep exploring until one or more mines is successful: on the average, it costs $20 million per successful mine, but since the success is worth $100 million, it's clear that in this game the investor with capital can't lose. Once again folk wisdom provides a nutshell explanation: "It takes money to make money."

Furthermore, people's strategies should not be judged unless we know their goals. The man with only one million dollars may feel that it isn't enough, and that unless he can build it into a much larger fortune he would be just as happy to be penniless. In that case he might just as well go ahead and buy his gold mine. In the imaginary game described, he might be the one person in twenty who will win. Most of us find it hard to understand how one could be unhappy with "only" a million dollars, so let's consider the exact same situation at the other end of the economic scale:

A contest based on pure luck of the draw has a single prize of $100,000. If ten million people are expected to enter the contest, an individual's chance of winning is 1 in 10 million. Another way of putting this is that the individual would, on the average, have to enter ten million such contests in order to expect to win once. Winning $100,000 after entering ten million contests represents an average gain of 1¢ per contest. Since it costs far more than this to mail each entry, everybody loses on the average. Yet this doesn't mean that entering a contest is necessarily foolish. The price of a stamp won't buy anything worth having, so if it is my entire fortune, I might as well spend it on my infinitesimal chance of winning the contest. This, of course, is why the numbers rackets can make so much money from the poor.

In real life we don't know the odds as well as we did in these cooked-up examples, but they should help to explain that what we can expect as an average return from an investment does not by itself determine whether an investment is good or bad.

In a chance situation the amount of capital we have influences the kind of strategy we should pursue. Lack of capital is what makes us buy insurance. The insurance company collects premiums that are based on the probability of having to pay benefits, plus some additional charge for expenses and some profit. On the average, the policyholder loses because of this additional charge. Where the policyholder gains is in the area of large but rare losses. We buy insurance to cover losses that are too heavy to be covered by our own resources, but that are also rare enough that we can afford the premiums. We should not buy insurance to cover losses that our capital can easily take care of, since then we're needlessly helping the insurance company to pay its expenses.

18
When Should We Take Chances?

In his day Knute Rockne was the nation's most successful football coach, so it was to be expected that coaches around the country would imitate his coaching style. His style was very conservative and consisted of playing safe, straight football and capitalizing on the opponents' errors. This strategy was ideally suited to his teams, which usually had the most strength and talent.

The lesson that other coaches should have learned from Rockne's success is that conservative strategies lead to success for the better team. What many of them thought they learned was that conservative football is winning football. College football became a dull, low-scoring game in which there was a lot of punting, frequently on third down. Then professional football began trying to attract customers by taking more chances — that is, by throwing passes more often, even from disadvantageous positions, and by using other plays that had potentially large payoffs if they worked but were costly if they didn't. This behavior not only attracted customers, but also showed the coaches of inferior teams that taking chances sometimes helped them to win.

Games consist of combinations of luck and skill. If you are better than your opponent your strategy should consist of eliminating the effect of luck as much as possible. It follows that if your opponent is better than you, nothing is going to make you win most of the time, but in order to win as often as possible you should do what you can to increase the role of luck.

How to Tell the Liars from the Statisticians

High school students often ask how much guessing they should do in taking a college entrance test. These tests are usually so graded that a wrong answer is penalized more than an unanswered question — enough more so that a student who makes nothing but random guesses will receive a zero, on the average. For a student who is sure of the incorrectness of at least one of the multiple choice answers given for a question, a guess among the remaining ones is profitable, on the average. However, anyone reasonably confident of doing well enough to get into the college of his or her choice should minimize the role of luck by doing very little guessing. Students who think they need some help from luck should obviously do more guessing. (This is an abstract mathematical approach to the subject. It neglects the question as to whether anyone is wise to try to get into a school when he or she has to have luck to make it.)

Some readers will take exception to the idea that there is so much luck in our lives. It is true that if a student knows "everything" that might be included on a test then luck is not a factor. But if a student knows only 70% of the material then it is luck (in that it is out of the student's control) whether 60, 70, or 80% of the test concerns the 70% that the student knows. If this student needs a 60% mark (for whatever his or her goal may be), he or she should not do much guessing; but if 80% is the goal, guessing is definitely in order.

A golfer often has to make a decision in a case like this: the green can be reached in one stroke by going over some tall trees, but it will take two strokes to go around. Going around is "safe" because it almost never takes more than two strokes, but going over the trees is a risk because if the ball hits a tree several additional strokes may be required to get out of the woods. Golfers know that near the end of a tournament the leader should play it safe, while those who are trailing by a little should take the chance, provided that the principal objective is to be first. But there is also money for second and lower places, and there are many times when these decisions come up early in the tournament, so the right choice is not usually obvious. The professional golfer who is good enough to make the total money he or she wants by playing it safe should probably do so most of the time, but the golfer who is not quite so good will have to take more chances in the hope of sometimes winning more than his or her talent deserves.

In commenting on other people's strategies in various activities we tend to use words such as "conservative," "aggressive," "unimaginative," or "dangerous," usually with the implication that a certain kind of behavior is always good or always bad. One of the things we should learn from thinking about probability is to ignore those who try to sell us on the idea that one particular way of playing the game is always best. Different circumstances, including the odds that vary from one situation to another, call for different kinds of reactions. The good strategist learns to overcome the common tendency to fall into a rut of consistent conservatism or consistent overaggressiveness.

19
Cost Accounting

A major league umpire, explaining his claim of never having made a mistake, once said, "Some balls are fair and some are foul, but they're never either one until I call 'em." In other words, the umpire is right by definition, and the rule book probably supports this. The prevalence of this condition in other areas is not widely known. Take accounting, for example.

Most people do not understand accounting, do not like it, and are more than happy to limit their contact with it to the famous "bottom line." As a result they force themselves to take the accountant's word on faith, and great is their disillusionment when external evidence shows that the accountant's word is human, and hence fallible.

What should be the postage rate for "occupant" mail as opposed to first-class addressed letters? Most people think they've answered this question when they say it should depend entirely on the relative costs of the two services. Asked how to determine the relative costs, they answer that that is a problem for the accountants, and they will believe whatever the accountants say. If, say, for political reasons, two different accountants are summoned to solve the problem, they will probably obtain different answers, and disillusion may begin to set in. The reason that two accountants can get different answers is that there isn't any answer until an accountant declares one. For instance, consider the salary of a postmaster of a large city. How much of his or her salary is a cost due to "occupant" mail, and how much should be charged to first-class letters? You can prorate his or her cost to the two sources according to volume of mail measured by weight, by number of pieces, by time spent, or by any number of other factors that an insider could probably come up with. None of these methods produces a "correct" answer in the sense that 4 is the correct answer to $2 + 2$. Each produces a different answer that has some degree of justification.

Figuring profits is another area with the same uncertainties. Those with the "bottom line" mentality like to pretend that profit is profit and that there is only one true and honest answer in every case. Profit is computed over a time period and the business in question is usually operating at full speed at the beginning and at the end of the period. We have only to look at the problem of evaluating inventory to see that profit depends on definition. We had some inventory at the beginning of the period and a different amount at the end. Assuming that this inventory is measured in dollars, should it be dollars that the inventory actually cost, or dollars (replacement value) that the inventory is currently worth? In a period of inflation, these can differ considerably.

There are laws governing the taxation of profits, and so there are also laws governing methods of accounting. These laws cannot reduce accounting to arithmetic, and they must leave to the accountant a certain amount of freedom to decide how to draw his or her conclusions. Since this freedom sometimes leads to disagreement, bystanders have been known to use the term "creative accounting" in a derogatory sense, trying to imply that where accounting is honest it will be pedestrian and it will have unique answers. It is true that complexities of the real world permit accountants to become effective data pushers if they so desire, but a user of the "bottom line" can sometimes avoid being a pushover by trying to understand how the accountants arrived at it.

20
Bridges, Underpasses, and Corporate Profits

A bridge builder wants to be very sure that the strength of a girder exceeds the load that the girder will have to support. A truck driver wants to be sure that the clearance of an underpass exceeds the height of the rig. A corporate president wants to be sure that gross income exceeds total operating costs. A statistician (or mathematician) considers that they all have the same mathematical problem, though this might surprise them. Each can solve his or her specific problem without thinking of it mathematically, but the mathematical approach can lead to greater understanding, especially of how similar other people's problems can be to our own.

With engineering experience and knowledge of physics, the bridge builder can estimate the strength of a given girder design, but there is always some uncertainty involved, since even supposedly identical materials vary somewhat in strength. The builder can estimate the greatest load that the girder will have to support, but this also is subject to uncertainty, since it depends, for example, on future traffic patterns and truck weights. The builder counteracts the uncertainty by adding a safety factor, that is, by choosing a girder that has greater strength than is expected to be necessary, enough so that unexpected variations in load will not cause the bridge to drop into the river below.

The truck driver whose rig is 13 feet 11 inches high, say, can sail right through a place marked "clearance 15 feet." But if the clearance is 14 feet, even though this gives an inch to spare, the truck may be seen creeping through at a very slow speed. The reason is that in the real world the clearance is subject to uncertainties, such as might be caused by a buckling in the road surface or an accumulation of dirt, and the rig height is also a bit unpredictable, since it changes a bit with load and tire inflation, for example. The driver knows that when the clearance and the rig height are close, these irregularities may combine to make the underpass hard to go through, and so slow driving is resorted to because it tends to reduce the effect of the irregularities.

The corporate president needs to show a profit, and so the gross sales must exceed the costs. Neither of these is exactly predictable, so for the sake of job security, not to mention corporate survival, the president wants to make sure that gross sales exceed costs. To this end, profits must be greater, as a percentage of sales, than the error in prediction of sales and costs. Building in a safety factor for profit as a percentage of sales serves the same purpose for the corporation that a safety factor does for the bridge builder.

How to Tell the Liars from the Statisticians

The statistician sees all of these problems as the same, saying simply that there is a quantity X (strength, clearance, or gross income) which may have small fluctuations, and there is also a quantity Y (load, rig height, or total cost) which may have small fluctuations, and that it is very, very important for X to be bigger than Y in spite of the fluctuations.

Bridge builders, truck drivers, and others have been known to accuse corporations of greed when the latter make a good return on investment but complain about their small profit as a percent of sales. The critics might be more forgiving if they would take the statistician's view that the three problems we've talked about are one and the same.

21

Are You Average Enough?

Sometimes it seems as though the world actually resents our departures from the norm. Do you bump your head getting in and out of cars? The car salesman will give you a look that suggests that people over 6 feet tall are freaks and should expect such trouble. Do your feet fail to reach the floor when you sit in a chair? The furniture salesman's unconcerned shrug can make you feel guilty and inferior for being so short. Is your malady different from "what's going around"? The doctor's skeptical stare can make you feel like a hypochondriac. Economists repeatedly assure us that inflation is not bad, because incomes rise along with prices; what they mean, of course, is that average incomes rise, and people with nonaverage incomes are ignored.

Manufacturers and salespeople have an ongoing love affair with the "average person," inspired by the belief that anything that pleases this mythical being can be sold in the tens of millions. Houses, cars, and furniture are designed on the assumption that everyone is 5 feet 6 inches tall, or whatever someone thinks the average is. During World War II some nonaverage people received a jolt when a bureaucratic decree outlawed the use of leather in shoes larger than size 12. Peacetime conditions are better, but if your shoe size is slightly odd, you don't need to be told that your choice of footwear is severely limited.

Mass production, understandably, aims at the average. What is a lit-

tle hard to understand is that now that we have computerized production and distribution, producers are giving us even more standardization instead of going in the opposite direction which computers might allow them to pursue. In market after market, any number of companies compete desperately for a share of the big middle, or average, group of consumers when they could have the nonmiddle market for the taking, and still have millions of customers. Can it be that we have too many average thinkers making our important decisions and programming our computers?

Market analysis helps the manufacturer to know what the consumer wants. Properly used it is beneficial to both parties. Overused, it can be both self-fulfilling and self-destructive. For example, if analysis shows that people like horror films, producers will saturate the market with them. Subsequent analysis will continue to show that youngsters are very fond of them (as they seem to be at this writing), possibly because they don't remember anything else. In this way a prophecy of continued success for horror films may be self-fulfilling. Then, if history is any indicator, an alternative will come along, moviegoers will observe that they have become bored with the horror genre, and the buckets of blood will go suddenly down the drain. In this way the prophecy can be self-destructive. (At least there is always that hope.)

22

Infatuation with Averages

Membership in almost any profession carries with it various rewards, for which members must pay a price that includes having to laugh at weak jokes about themselves. In the case of statisticians these jokes usually take the form of references to the statistician who had 2.3 children and 1.7 divorces, or, even more hilariously, to the statistician who placed his head in the oven and his feet in the refrigerator and reported being comfortable on the average.

The basis of such jokes about statisticians is the implication that they are infatuated with averages. Actually it's other people who are infatuated with averages and statisticians who are forever trying to set them straight. Mass production is only one aspect of this phenomenon; we know that excessive love of averages must go farther back or we wouldn't have inherited from the ancient Greeks the well-known legend of Procrustes. Procrustes engaged in the uncommendable practice of fitting his guests to his bed by stretching them on a rack or lopping off parts, as the guest's dimensions required.

Procrustean bed thinking, on a more abstract and less bloody plane, is still around to cause many of our difficulties. People faced with a situation involving variability, or probability, usually don't know the mathematics needed to handle it, so they try to circumvent the problem of variability by assuming that the average prevails uniformly. This unrealistic assumption is often made by people who pride themselves on their practicality, and sometimes it works, but frequently it does not. One of the places where it does not is in the problems treated by what mathematicians call "queuing theory." This is the term for the mathematical theory of waiting lines, and it came from the British, who form queues instead of standing in line.

The waiting lines in a bank give us a simple example that everyone has experienced. Suppose that customers enter a small bank on the average of one per minute and each has a transaction that will require one minute, on the average, to complete. How many tellers are needed? In the attempt to answer this question, it may be tempting to assume that the average prevails, that is, that a new customer enters the bank each minute on the dot and takes exactly a minute to be served. If this were really the case, then obviously one teller would be able to handle the demand exactly. Alas, in real life the customers arrive randomly and take different amounts of time to be served. In this more complicated situation, even if customers still average one arrival per minute and require an average of one minute of service, it is a mathematical fact that one teller will be unable to cope. Actually, with just one teller in the real case, the average waiting line will build up indefinitely. A part of the explanation for this is that with variable arrival and service times, the teller will sometimes have nothing to do, and the productivity lost during these idle times can never be regained. (A fuller explanation is more mathematical than anything we want to do here. It can be found in textbooks on queuing theory.)

23

The Law of Averages, Compulsive Gambling, and the Rubber Band Theory

Psychologists like to explain the gambling habit as a compulsive desire to lose. We may some day find that the psychologists have a compulsive desire to contrive inverted explanations.

The simplest explanation of the gambling habit is that it starts with a desire to get something for nothing and continues because the gambler thinks the law of averages will eventually enable him to recoup his losses.

Losers usually think that the law of averages acts like a rubber band. They think of losing as a stretching of the band, so that the more one loses (or stretches the band), the greater becomes the force to restore the losses (or snap the band back). It is not like this. The more you lose, the longer it will probably take to win it back, until eventually the required time may be more than you have. It is a temptation to point out that rubber bands eventually break under the strain; the analogy may not be exact, but if it helps you to remember that your fortune may vanish before your luck begins to turn, so much the better.

The law of averages says, roughly speaking, that luck evens out in the long run. There are various mathematical ways of stating this law, and when stated mathematically it can also be proved mathematically, but what interests us here is what people think it says. The long run is so often longer than they think.

Suppose you're tossing coins with a friend who asks for a starting advantage of 20 wins. Naturally you take a dim view of this and respond that nobody can be expected to overcome such a handicap in an even game. On the other hand, if your friend achieves a 20-win advantage from luck in actual tossing, most people seem to believe that your luck will quickly turn and wipe out this advantage. Clearly, however, the coins are not able to tell the difference between a 20-win handicap that has been given and one that has been achieved by tossing. No matter which case prevails, the 20-win handicap will not be quickly eradicated. The loser who expects quick reversal of fortune is almost always disappointed and will often turn to superstitious explanations of what has happened.

If two people toss perfect coins, the percentage of times that one person wins will almost certainly come closer and closer to 50% as more and more tosses are made. In this sense, what people believe about the law of averages is true. What people are not aware of is that the "law" allows occasional runs of "good luck." For example, if two coin tossers each have a stake of $1000 and they toss for a dollar a throw, eventually one of them will find his or her entire stake wiped out, and will probably ask, "Where was the law of averages when I needed it?"

Actually, it takes about a million tosses, on the average, before one participant loses the entire $1000. Note how the interpretation varies according to the gambler's wealth. A multimillionaire who has gambled a million tosses and lost "only" a thousand dollars is apt to observe that the law of averages has been confirmed; he has won 49.95% of the tosses, as near to half as anyone could reasonably expect, and he may consider $1000 little enough to pay for the entertainment. The gambler whose $1000 stake was his or her entire fortune, however, sees only that it is gone, and blames hard luck.

Whether or not you can be philosophical about losing depends on how much you have left.

24
A Rational View of Luck

There are those who claim to be lucky, those who say they are unlucky, and those who say they don't believe in luck. Almost everyone has something to say about luck, and few subjects receive more comment by persons having less correct information. In spite of all the mystique and confusion surrounding this word, it is nevertheless possible to take a rational view of it. Since luck is a concept associated with probability, this is a good time and place to discuss a rational approach.

Those who say they don't believe in luck are often simply those whose luck up until now has been largely good — as any sports fan knows, athletes like to take full credit for winning streaks, and so do people in all other walks of life. Some who claim not to believe in luck really mean that anyone who has sufficient courage, grit, foresight, and determination can overcome such bad luck as may come along. Numbered among these are the centenarians, who usually have a recipe for living a hundred years: many credit their habits of total abstinence, while the remainder recommend a daily dose of bourbon and cigars.

Others who don't believe in luck are really saying that they believe everyone's luck evens out. Few such believers were born in a ghetto or paralyzed by polio at age 3. As in the case of gamblers, our fortune may vanish before the good luck begins; life is a gamble, and the bad luck of being killed by a truck at age 20 can't be balanced by any amount of good luck that might have been expected to occur later.

A person claiming to be "lucky" is making perfectly good sense if by this it is meant that he or she has had good luck up to now. If the claim extends to the future by implying a greater chance of winning a lottery than the rest of us have, then it becomes a superstitious claim, and this is the kind of luck that many are speaking of when they say they don't believe in luck.

Those who so wish may go on thinking of luck as being like a fairy godmother that hovers around certain people and prevents them from being overtaken by common disasters, or like an evil demon that does the opposite. It seems undesirable, however, to waste a perfectly good word on a superstition, so I propose that a rational definition is just this: luck is the effect of circumstances over which we have no control.

To illustrate the various views of luck and how they conflict, let's suppose that my boss has entrusted me with some money to take to a bank for deposit. While I am on my way, a motorcycle goes out of control and knocks me down. While I am lying unconscious in the street, someone steals the money. Later I report this to the boss, summarizing by saying that it was bad luck. The boss says she doesn't believe in luck and fires me. She probably thinks she is using a rational definition of luck, but she isn't. If she admits that the circumstances were beyond my control, then firing me doesn't make any sense, but simply shows that she is going to look for an employee with a more potent fairy godmother.

Even some of those who try to be rational about luck end up becoming superstitious because they observe events that seem to defy what they think are the laws of probability, so that they seek another explanation. Often what they should do instead is to learn a little more about probability, particularly about the unusual events that are likely to occur to people who keep giving them the chance to happen. If you and I toss coins, the chances of your winning the first ten times are very, very small, 1 in 1024. If you were a stranger and won the first ten coin tosses from me, I would justifiably find an excuse to terminate the game. On the other hand, if you were a friend and we tossed for a drink every day after work, it is quite likely that you would win ten times in a row *at some time* in a four- or five-year period. Knowing this and expecting it, I would probably still grumble when it happened; not expecting it, I might invent a mythical creature called Lady Luck to explain it.

Mathematicians like to talk about coin-tossing games, because they can figure the odds. At something more complex, like the game of bridge, the odds may be impossible to compute. For example, in bridge I can't compute the chances that one pair will score 10,000 points before an equally capable pair scores any. I can say, though, that (leaving out the occasional points for honors) only one side scores points on a given hand. Then if contesting pairs are evenly matched, each has half a chance of winning points on each deal, just as for the coin toss. A pair who wins ten hands in a row is having a fabulous game, and the 1023-to-1 odds against this are so high that the lucky winners will be sure that recent efforts to improve their game have been appropriately rewarded. However, if you play twenty hands an evening, twice a week, don't be surprised if one side wins ten hands in a row *at some time* during the first year; and if it happens against you, don't be surprised if the score takes a very long time to even out again. So before you become discouraged about your skill, or about your luck if that is your belief, remember that I warned you!

25

Luck and Skill Combined
The Regression Fallacy and
the Sophomore Jinx

Once, I'm told, a certain judge decided that circumstances called for him to establish a line separating all games into two categories — games of chance and games of skill. He extricated himself from this self-imposed trap, with something less than Solomonic wisdom, by decreeing that a game is one of chance if the player cannot succeed by skill more than half the time. The judge was "right," of course, just as the umpire is always right, even though not far away the living legend, Willie Mays, one of the most skillful baseball hitters of all time, was still demonstrating that even the best hitters do not succeed half the time.

Most games, including all those we call sports, combine skill and luck in varying degrees (luck meaning the effect of circumstances that are beyond the players' control). Most players of these games have no trouble comprehending this even though their grasp of the elements of abstract probability may otherwise be very tenuous indeed. Lengthy arguments can develop about the degree to which luck enters in.

If you look at the ten individual leaders in any sport for a given year, based on any given statistic, whether it is average golf score or number of touchdown passes, and then look at these same individuals the following year, you will usually see that most of them have not done so well. Similarly, look at the ten worst, and you will usually find that, among those who are still around, most will have improved. Some have concluded from this that there is a general regression toward mediocrity, but this is false reasoning and is often referred to as the "regression fallacy." Looking at the whole group of athletes will show that there is no regression toward mediocrity — next year there will still be extremes of good and bad, but they will not be the same athletes who were this year's extremes.

The ten best in a field (sports, business, whatever) are there because of a combination of skill and luck. To deny the effect of luck is to claim that there are no uncontrolled factors. Furthermore, when the competition is fairly equal in skill, this means that outcomes are determined by luck to a high degree. A golfer who wins a tournament must play very skillfully, but the invisible winds aloft or a late onslaught of rain affect some more than others, and that is why next week's winner will probably be someone else.

The well-known "sophomore jinx" is an illustration of the same effect at work. Why does the sensational rookie so often perform disappointingly the second year? There are various possible reasons, one of which is a statistical one. Some athletes are exceptionally good, and in a given period of time some athletes have good luck, and those who combine the two are standouts. If the luck feature was strong, the standout can't expect a repeat performance. Even if his or her second-year performance is still good, if it isn't as good as the year before, the sophomore jinx will be mentioned.

It is doubtful if the human race will ever stop being confused in this area. One reason is that no matter how important skill may be in a given game, if two contestants are exactly equal in every aspect of skill the outcome will be entirely a matter of luck. One year the Pittsburgh Pirates won the National League pennant and a reporter asked Tom Seaver (then a pitcher with the losing Mets) if he thought the Pirates were lucky to win. Seaver, who is more thoughtful than a lot of people, replied that in a league with such evenly matched teams you have to have luck to win. Inevitably, the headline read, "Seaver Says Pirates Lucky to Win."

26
Luck and Dependence

What are your chances of being in an auto accident and being struck by lightning on the same day? Let's suppose each of these accidents happens at the rate of one every thousand days, just to use round numbers and keep the waters from being muddied by too much arithmetic. Then I hear the mathematicians among you saying that the chance of experiencing both disasters on the same day is found by multiplying, so that the answer is one in a million. Right? Well, yes and no.

A sudden thunderstorm slicks the road. You skid off the road, hitting an isolated tree, and must get out of your car because it is about to burn up. This exposes you to the lightning, and suddenly your chance of being struck has gone up a great deal, so that it would not be inconceivably bad luck to have both disasters happen on the same day. Similarly, if the chance of needing an emergency operation on a given day is one in a thousand, what is the chance that you will need the operation *and* have an auto accident? Again you can't automatically give the one-in-a-million answer, because the need for emergency surgery may induce your spouse to drive you through town at 70 miles per hour, causing your probability of having an auto accident to soar. A mathematician would say that in this case the probabilities are not independent.

Mathematical results are based on assumptions that certain conditions exist. Results are usually more interesting than assumptions, so peo-

ple tend to forget the latter, which are nevertheless crucial to validity. The law of multiplication of probabilities rests on an assumption of independence, that is, that the outcome of the first event does not influence the outcome of the second. When there is dependence, as is often the case in real life, then the simple multiplication rule doesn't work. This partially explains why every now and then we have "one of those days" when everything goes wrong. Often a retrospective look at such a day will show that some of the things that went wrong were dependent — that is, they were in some way connected, or had a common cause, and that we should have been prepared for them to happen simultaneously.

27
Confidence and Dependence

A fairly good golfer who hits reasonably straight drives may get a few bad bounces that cause several drives to end up in the rough. Suddenly the fairways seem to have shrunk to a width of about 10 feet. The golfer has lost his or her confidence, and for a while there will be fewer drives into the fairway, until some lessons or some good luck restore this mystical but very real and important feeling. Similarly, a batter who has hit several hard shots straight into the hands of fielders begins to feel that there are far more than nine people out there, while at the same time the ball to be hit seems to have lost much of its diameter. If a game depends on both luck and skill, and if the bad luck starts a losing streak, loss of confidence can prolong it. Winning streaks work in the opposite way, with a little good luck promoting confidence. Sometimes, however, the athlete can become confident to the point of feeling invincible, and such overconfidence can cause a winning streak to come to an abrupt end. In sports like golf, baseball, and tennis the player starts being overconfident when accuracy becomes so good that greed asks for a little more power, and this may play havoc with the smooth swing that generated the confidence in the first place.

We've seen that streaks (of wins, losses, successes, or failures of any kind) can be the result of pure chance, and that they can be enhanced by various degrees of what statisticians call dependence. Dependence, in turn, is often increased by confidence or lack thereof. Another source of streaks is redefinition. Sports reporters talk about streaks constantly, probably grossly overestimating the public's interest in most of them. If a streak is broken, the reporter, needing something to fill the airwaves or his or her column, redefines the streak so as to continue its life. So if a team loses after winning twenty games in a row, and if the loss was out of town, the reporter begins to count the number of successive home games won by the team. Thus was born Chris Evert Lloyd's record of 64 straight wins on clay courts, and, more trivially, the news passed on recently to me that the Chicago Cubs had lost 17 games in a row on artificial turf. Like the proverbial man-bites-dog story, the ultimate story may eventually be, "There are no streaks of interest alive today."

The effect of confidence on athletic performance is well known, but it has received little recognition in nonphysical activities. In the game of bridge, for example, confidence can promote success or failure much as it does in sports. It seems to affect the decision process. A little bad luck can render a player indecisive, and cause him or her to assess chances in terms of recent events instead of in terms of overall experience and known odds. A hand that has a better than half a chance of producing a small slam should be so bid, but an inexperienced player may be afraid to bid it if recent attempts at aggressiveness have failed. Even if the proper contract has been reached, such a player may play the hand badly, as the unconfident mind tends to wander down alleys of irrelevant speculation while it should be thinking positive thoughts about upcoming decisions.

If there is any truth in all this amateur psychology, it must follow that winning and losing streaks have similar effects on the confidence or lack thereof among people making business decisions, and that a degree of confidence (but not too much) may be just as necessary there as in sports or card games.

28
Sports "Form" and Consistency

If Atlanta beats Baltimore most of the time, and Baltimore beats Cincinnati most of the time, then it is usually assumed that Atlanta will beat Cincinnati most of the time. If Atlanta has never played Cincinnati, this "form" (racetrack word for summary of past performance) is the only thing the experts have to go on in predicting the winner. If Atlanta goes against "form" and loses to Cincinnati, the prophets will excuse their inaccuracy by saying that in one game any team can win. If Atlanta persists in losing to Cincinnati they may decide that Atlanta has some psychological hang-up about this particular team. Or they may try to convince us of their expertise with more complicated explanations based on relative strengths of defensive and offensive teams, if the sport is football, or of pitching and hitting if the sport is baseball.

Any or all of these explanations for going against form may be valid, but it may also be that the answer is a simple statistical one based on consistency. We all know that average performance is only one aspect of the capability of a team or of an individual, and that consistency affects outcomes, even over the long haul, and even when averages are all the same.

Imagine three computer-controlled mechanical golfers that are designed to score as follows:

A shoots 72 all of the time.
B shoots 69 one time out of four.
B shoots 73 three times out of four.
C shoots 70 half the time.
C shoots 74 half the time.

When one of these "golfers" plays another head-to-head, there is no effect of psychology and there is no effect of the opponent on the play. Each golfer averages 72 over the long haul. Golfer A will clearly beat golfer B three times out of four (that is, whenever B shoots 73.) Golfer B will beat C whenever B shoots 69 (one-fourth of the time) plus half of the rest of the time; this works out to a total of 62.5% of the time. If golfer A had never met golfer C, "form" would tell us that A will beat C fairly consistently. Yet the numbers show that A will beat C if and only if C shoots a 74, which is exactly half the time.

In short, A beats B most of the time, B beats C most of the time, and C comes back and plays A exactly even. In average scores there is no difference among them at all. A's greater consistency is an advantage against B, but not against C. The odds are indeed odd.

29
Who Is Unemployed?

Mazetti took a year off to work, but now she is back at college, studying to be a lawyer. Williams is a professional baseball player who is taking it easy this winter. Jackson is a writer, has just sold her latest book, and is starting to think about the next one. Hernandez has been working part-time since graduating from college, and is looking for a full-time job in advertising. Jones recently went back to work after getting her children into school, but six weeks later her company started an austerity program and she was laid off. Robinson has retired but takes on consulting jobs and has not had one lately. Malone is an actress who has a job starting in three weeks, but is not presently working. Gregory says he would like to have a job but he really isn't looking. Patrick is looking for a job in personnel, but she didn't do very well in school and the jobs that have been offered her are not to her liking.

How many of these people are unemployed?

To most of us the difference between having a job and not having one is a difference as great as night and day. If a full-time worker supports himself or herself, and perhaps also has some dependents, loss of the job is a change of status that doesn't require elaborate measurement instruments to detect. But there are a lot of people who exist in a twilight area between employment and unemployment, so that it's hard to tell in which group they belong.

The Bureau of Labor Statistics, being charged with the task of measuring general unemployment, has to decide which of these cases to count. These inbetween people are not numerous compared to the population at large, but they are numerous compared to the group of unemployed. Arbitrary decisions must be made about who is unemployed and who is not. The inbetween people are also harder to enumerate, since they may not be as much in evidence as the truly employed or the truly unemployed.

The unemployment index is an important guide in the analysis of the overall economy, but, like all statistics involving people, it is not the precise number we would like it to be, nor always as full of meaning as we may think it is.

30
Cost of Living

When the government reports that the inflation rate is 12%, citizens like to think of this as a firm figure on which they can base decisions. Unfortunately it isn't.

The price of eggs and the price of yachts do not move upward at the same rate. Most people buy eggs, but very few buy yachts. How much should each one count when the government tries to set a single index as a measure of the inflation rate? Many would be happy to leave out yachts altogether, but of course the problem extends to more common things, such as the cost of a home. To those who already own their home the changing cost of houses may be of little interest. To those who are looking forward to their first owned home, its rising price is a major factor in inflation. A statistic that compromises between these two views will fit almost no one.

Official estimates of the inflation rate are desirable for various purposes, such as the determination of increases in social security benefits and salaries of civil service workers. On the other hand, these estimates may be undesirable, since they provide a means of virtual price fixing if everyone who is in a position to set prices simply raises them in accordance with the latest index.

A business magazine at hand complains that a sudden drop in the consumer price index is partly due to a new allowance for discounts on autos at the end of the year. It is strongly implied that the reason for mak-

ing this change was a political one, aimed at showing that the administration has brought inflation under control. In the next paragraph the editorial points out that the consumer price index suffers from some deficiencies and should be changed. In short, if the procedure for calculating the index is never changed, that is a sign of inability to carry out needed improvements; but if it *is* modified, that is a sign of political maneuvering!

The price of eggs may be a hard fact, but the consumer price index, like any other capsule description of human affairs, is only a rough attempt to summarize a lot of information in one number.

31
Trade-Offs

When a drug turns out to have serious unforeseen side effects, headlines occur. People say, indignantly, "They should have checked it out first," and litigation grows by leaps and bounds.

Certainly when the side effects appear, the headlines are justified, and strong measures may also be, but the other reactions often are not. The fact is that drugs are tested, invariably. The investigator who has observed some beneficial results on a few patients has a dilemma: should the results be announced, or should experiments continue until they are conclusive? If results are announced too soon, many people may gamble their lives on the drug before it is found to be worthless or even harmful. On the other hand, if investigation continues for another year or so and the drug is found to be beneficial, some people who might have been saved by the drug may die in the meantime. Some of the same reporters and editors who complain that a drug should have been more thoroughly tested will also complain when they hear that a possible cancer cure has been kept off the market for more experimentation.

If the investigator announces his results too soon, and it turns out that the drug really did no good, this is called by statisticians a Type I error. If he decides to continue investigating when he really ought to stop and announce a discovery, this is a Type II error. People who believe life to be made up of simple yes–no decisions don't like to hear about these two kinds of error, but the fact is that they pervade the whole process of living. The only way to reduce one kind of error, for a given expenditure of time and money, is usually to increase the other kind. Scientists of all kinds are suffering from having built up an image of omniscience. It may make sense to sue an investigator for being careless or dishonest; it does not make sense to sue one for not being all-knowing.

32
False Alarms
Or Type I Errors

As I write this, my newspaper is carrying a headline that reads, "Hundreds of False Nuclear Attacks Reported." This kind of thing tends to scare readers, probably intentionally, but false alarms are a fundamental part of any kind of detection.

A good example is the equipment used at airports to detect arms being smuggled aboard planes. This equipment is not perfect and sometimes it will not be sensitive enough to detect a particular weapon. Failing to detect a weapon when one is there is called a Type II error. Most devices can be made more or less sensitive, but for a given design, increasing sensitivity usually also increases the false alarm rate. A false alarm, or Type I error, consists in announcing the presence of a weapon when there is none.

Those who constantly wonder why life is not more nearly perfect will ask why we can't reduce both types of error. Actually, of course, research may well be able to accomplish this. Often, however, the improved equipment turns out to have some other defect, the most common one being that it is more expensive.

All methods of detection are subject to these two types of error, from smoke alarms and burglar alarms to methods of detecting cancer or tuberculosis. If an x-ray procedure is to be able to detect nearly all cases of lung cancer, it will produce a certain number of false alarms. Our best protection against false alarms is to have a backup procedure for checking things out before we take any unjustified action.

Statisticians introduced the concept of errors of Types I and II to describe and solve certain problems, and there seems to be no end to the situations in all areas of life that can be clarified by this idea. Nothing can be farther from statistics than the Book of Common Prayer, wherein is stated, "We have left undone those things which we ought to have done; and we have done those things which we ought not to have done." When the great day comes that statistical education is available to all, we'll be able to shorten this to, "We have committed both Type I and Type II errors," and everyone will know what is meant.

33
Liberals and Conservatives

A good administrator of a welfare program tries to distribute benefits so as to avoid neglecting those who deserve them, while at the same time not squandering available funds on recipients who don't need them. Wasting money on those who don't need it is responding to a false alarm, or Type I error. Allowing someone to starve who should get welfare benefits is a Type II error.

Our judicial system declares a person innocent until proven guilty beyond any reasonable doubt. Declaring innocent people guilty is a false alarm, or Type I error, while declaring guilty people innocent is a Type II error.

Discussions of subjects of this sort are greatly simplified when the discussants both use and understand the terminology of Type I and II errors. In the more usual type of emotion-laden argument, articulate participants can create the impression that their position does not produce errors. It's much harder to engage in such pretense after the errors have been brought out into the open and specifically defined.

Liberals and conservatives justify their positions with much detailed analysis, but when the smoke clears it usually comes down to this: one side is telling us how diligent they are at avoiding Type I errors, while the other is going all out to avoid Type II errors. On different issues they take different sides. When money is being given out, liberals try to avoid Type II errors while conservatives aim at reducing type I errors. When punishment is being administered, their roles may be reversed.

In many cases a proposed law is desired by everyone, liberal and conservative alike. Controversy then arises only over the way the law is to be written and, later, interpreted. In a general sense, the liberal position usually is to write the law so as to avoid inaction, that is, to avoid the Type II error of failing to act when action is called for; conservatives often fear more the Type I error, which is to take action when action ("government interference") is not needed.

One reason people take different positions on the two types of error is that they have different views on the trade-offs involved. Laws have side effects, just as do drugs. (A minimum-wage law has the avowed purpose of raising wages to a decent level for those whose skills are not great enough to demand such a level. But it also has the side effect that it will cause some workers to lose their jobs when their employers can't afford the higher amount.) Whether we are more afraid of inaction than of action, or vice versa, depends on how we view the importance of the side effects of action.

To some extent what is Type I and what is Type II is a matter of definition, and in none of these examples is either group of people right in principle, however loudly they may claim that they are. It's always a matter of degree as to how many errors of one type we're willing to tolerate in order to reduce the occurrence of errors of the other type. Only the extremists in either camp take the simple-minded position that there is only one type of error, and only the dreamers really believe that we can eliminate one type of error entirely without allowing the other to increase unreasonably.

34
One-Armed Consultants

At one point during the writing of this book, it occurred to me that I might be overusing the phrase "on the other hand" to the point of being tedious. This reminded me of the senator who became so tired of the phrase that he proposed to put an end to it by hiring only one-armed consultants.

Granted the senator was having his little joke, he was also somewhat serious, and what he was saying in effect was that he was tired of consultants who gave him the facts and left him to make the decisions. If this was the way he felt, I wonder what he thought his job was.

The fact is that making decisions is the most important part of most jobs. Sometimes we may yearn for an easy one for a change, but the easy decisions are the ones that we made long ago or presently with such ease that we don't count them in with the rest. It's always the hard ones that we're left with, and they're hard because they have a set of arguments on the one hand and an opposing set on the other. That's why an important branch of statistics is devoted to decision theory and why even a simple book like this one spends a lot of time on the subject of trade-offs.

Anyone who runs for the U.S. Senate expecting to have a nice, easy job making decisions on matters that don't have two sides must have led, up to that time, an extremely sheltered life. Decisions are generally based on information, and information usually includes data. It seems safe to say that if someone presents numbers to the senator (or even to you and me) so that they appear to lead to just one simple, inescapable conclusion, that person is very likely a data pusher whose motives should be examined.

35
The GGFTGN, the Bill
of Rights, and Other Things

Advertisers and politicians often like to assure us that they are maximizing our benefits while minimizing our costs. A moment's thought reveals that this combination of buzzwords doesn't make sense. Minimum costs occur when we don't spend anything, and this will not produce maximum benefits for anyone. Faced with this argument, the perpetrator usually says, "You know what I mean." But I don't, unless what he means is that he wants to take a position that can be moved around to suit the occasion.

The same argument applies to the GGFTGN, the "Greatest Good For The Greatest Number," that some believe is the foundation of democracy. They see our political system as providing maximum freedom with minimum infringement of rights. This is not what it does, because this is not even conceptually possible. Maximum freedom, like it or not, is pure anarchy. Minimum infringement of rights occurs only with constant policing to insure that no one interferes with the rights of anyone else, and such policing removes a great deal of freedom.

Generally, if two things have any common causes, it is not possible to maximize one of them while maximizing or minimizing the other. This is a mathematical fact. What you can do is to constrain one thing to be less than (or perhaps greater than) some limit, and then maximize the other thing within this constraint. Our Bill of Rights works this way by prescribing the limits beyond which we can't infringe on individuals' rights; within these limits we try to give people as much freedom as possible. Fortunately, the founding fathers didn't just say they wanted the GGFTGN, which would have left us in a state of hopeless confusion as to what our aims are. Instead, they said, in effect, that the majority would rule, but that there must be specific limits on what they can do to the minority.

Corporations have been known to state that their goal is to "maximize profits while minimizing costs." This again is not possible. Costs can be easily minimized by laying off all employees and closing shop; having done this one must face the fact that profits will not be maximized. Obvious? Apparently not to all. Many would have corporations try to maximize a number of different things at once. We can and should expect them to maximize profits within limitations on the amount of harm done to the environment, the employees, etc. We cannot ask them to minimize harm, since that would lead to the ultimate harm, namely, closing down.

An old joke summarized this problem. Remember the mother who gave her son two neckties? Whenever he wore one of them, she asked him why he didn't like the other one. Never pass up the chance to have your enemies committed to maximize two things — no matter where they put their effort, you will always be able to condemn them for not putting more of it elsewhere.

36
Things to Think about while Waiting to See the Doctor

Among the richest of our queuing experiences are those we have in the doctor's waiting room. All the requirements for congestion are met: random customer arrivals (the emergencies), variable lengths of treatment time, and often too many patients per doctor.

If the doctor could predict exactly how much time each patient would require, if all patients showed up on time, and if there were no emergencies, then a schedule could be drawn up that would require no waiting. Since none of these conditions prevails, the doctor tries to make a schedule that works reasonably well for everybody. Once again, the Greatest Good For The Greatest Number notion doesn't work, since what works well for the doctor doesn't necessarily work well for the patients, and vice versa.

In practice a doctor's schedule is arrived at empirically, as mathematicians like to say, or, to put it in simpler language, by muddling through. That is, if the schedule proves to be too tight, so that patients are still waiting long after the doctor expected to be on the golf course, the doctor may order the receptionist to schedule fewer patients per day. If the schedule is too loose, the doctor will suffer some idle moments during the day, and the opposite corrective action will be taken.

This is a classic trade-off situation, which many businesses now handle with "cost–benefit analysis," that is, by a mathematical balancing of costs and benefits. Such an analysis recognizes that different actions result in different benefits and different costs. We would like to take the action

that maximizes benefits per unit cost, but to learn what this is we have to assign dollar values to both costs and benefits. People often shy away from cost–benefit analysis because in real situations it usually isn't easy to decide how to make such an assignment.

Suppose, for example, that we set out to optimize a doctor's schedule from the point of view of society at large. Knowing that a loose schedule would cause the doctor to lose some productive time when no patient was available, we would assign a dollar value to the doctor's time. This would not be hard to do. What is hard is the assignment of a dollar value to patients' time, which would have to average over such diverse individuals as corporation presidents, salespeople, retired persons, and children.

Faced with the difficulty of assigning costs to variable quantities and to intangible quantities (such as the medical welfare of a patient), people often say that their particular problem defies cost–benefit analysis. But they will handle it somehow, if only by muddling through, and from the way they handle it we can generally infer what they really believe are the dollar costs. For example, if your doctor keeps an average of five people in the waiting room at all times in order to keep from ever having a nonproductive moment, then we can infer from that that the value assigned to your time is essentially $0 per hour.

37
Zero or Nothing?

One of my college professors liked to ask his classes, "What is the difference between zero and nothing?" After allowing the students several minutes of metaphysical and sophomoric discussion he would write "0" on the board and say, "That's zero." Then, with a flourish he would erase the zero explaining, "And that's nothing."

Years later I used to listen to a baseball announcer who with very little justification prided himself on the exactness of his language, particularly in dealing with rules, arithmetic, and the pronunciation of foreign names. In announcing batting averages, he avoided ever saying that a batter had an average of zero. Instead he would say of a batter who was hitless after several times at bat, "He has no average." He was wrong. A batter with one or more at bats and no hits does have an average, and that average is zero.

The announcer was right, however, in the case of the player who had not been at bat. Such a player has no average, and it is incorrect to say that he has an average of zero. Zero is a number, and should not be used as a catch-all of indeterminate quantities. As was noted in the discussion of patients in a waiting room, the overall value of their lost time is hard to estimate. The natural way to face up to this problem seems to be to run away from it (that is, to refuse to estimate the value). Unfortunately, not making an estimate often turns out to be equivalent to making an estimate of zero, which is not at all the same thing.

A very real example of this occurs in large companies where it is common practice to have employees in one division of the company sell their time, at cost, to another division or even to outside agencies. This cost is computed in various ways — usually as a salary-cost per unit time plus an overhead figure. Since these employees are usually professional people (not in production or sales) the value of their services is hard to estimate, so it is treated as if it were zero. To the accountants, this error of omission is of no concern, as it generally does not affect their objectives. Executives, however, who tend to use accounting figures to make decisions, may become victims of the confusion of zero with indeterminacy. For example, if an employee manages to sell only a few hours of his or her time per year, he/she will eventually be laid off; during those few hours the employee may have made suggestions worth millions of dollars to the company, but since this worth is hard to estimate the system ignores it (that is, estimates it at zero) and the employee appears to be of little value. This example is extreme and hypothetical to make a point, and if millions of dollars really resulted from one employee's suggestions, someone would probably step in and countermand the system. However, the bigger the company the harder it is to take such individual action. Furthermore, hundreds of thousands of more typical employees, caught in such a system, find it hard to take an interest in their personal productivity when the system always evaluates it at zero.

An interesting sidelight on this custom of our large companies is that they liken the custom to free enterprise. They argue that the value of an employee can be determined by the amount of her time she can sell, since the various company divisions (and outside agencies, perhaps) will use the employee's time only if it is considered to be worth the cost. Actually, it is nothing like free enterprise; the price of the employee's time is fixed by accounting and is out of her control. If she is better than the competition, there is no way she can turn this to her own or her division's advantage by charging a higher rate. On the other hand, if there are people outside the company, consultants, for example, who can do (or appear to be able to do) the work she does, she can't compete with them because they don't have to pay the overhead that she must pay. This overhead may or may not have anything to do with her actual costs, since it is usually a figure that is averaged over many employees and equipments. Nevertheless, this whole procedure is popular with executives, since as long as they can pretend that a free-market force is operating they can feel relieved of the duty of making such decisions as whether to use an in-house person or an outside service, or even whether to keep the in-house person at all.

How to Tell the Liars from the Statisticians

Accounting figures are a blend of facts and arbitrary procedures that are designed to facilitate the recording and communication of business transactions. Their usefulness in the decision process is sometimes grossly overestimated.

38
Quality Control and "Zero Defects"

When the walls of Jericho came down, someone on the inside probably said, "They don't make walls like they used to." The quality of our purchases is seldom as good as we would like, and we usually think we can remember when it used to be better. Fed up to here with products that go into a sharp decline soon after their warranties expire, we are prone to accuse manufacturers of engaging in "planned obsolescence."

The truth is that quality is usually obtained in a trade-off against price. Occasionally we benefit from a new idea that makes a product both better and cheaper, but such ideas are not commonplace. Even where quality is of supreme importance, there must be trade-offs. The owner of a pacemaker certainly doesn't want anyone skimping on the quality of that product, but he also doesn't want so much quality that he can't afford a pacemaker.

Scientific "quality control" has been on the industrial scene for only about a half a century, but it has helped to make our trade-offs much more beneficial to both producer and consumer. Its basic tenet is that before quality can be controlled it must be measured. A manufacturer of firecrackers needs to know how many "duds" are being produced, that is, what fraction are defective in that they will not explode. This fraction must be estimated from a sample, since any firecrackers tested are of course destroyed; making inferences from a sample means that we have to use statistical analysis to determine how large a sample is needed, how it is to be collected, and what we can say about the fraction that are defective in the entire output of firecrackers from what we see when we test a sample of the output. Clearly the manufacturers of firearms, bombs, and missiles have similar problems. For other products it may not be a "go — no go" situation in which something works or not, but one in which the product works better according to how close some measurement is to a given standard. Pistons for use in cars, for example, are manufactured so as to deviate very little from some desired diameter. No matter how good the manufacturing process is, there is always some slight variation in diameter from one piston to another. Usually we can measure more precisely than we can manufacture, so we can improve the output of pistons by measuring their diameters and discarding those that deviate too much from the standard. The more demanding we are in this respect, the better the ultimate car will be, but also the more expensive, since someone has to pay for the discards.

Before quality control began to be used, asking a manufacturer about his rate of production of defective units was a little like asking a politician his batting average on breaking promises. The answer was likely to

81

be as favorable to the responder as he thought he could get away with, since when no one is keeping score everyone can claim to be a winner. During World War II, when numerical methods began to be used to check the quality of military equipment, mention of possible failures was unpopular with generals, some of whom regarded failures as a form of insubordination that would not be tolerated. Eventually, however, it became accepted that in order to improve equipment reliability we have to start with the admission that it is not already perfect. If someone suggests a change to improve quality, we can't be sure it does so unless we measure the quality.

In recent years there has been an attempt to revive the simplistic approach by pushing a program labeled "Zero Defects." The idea is that defects are 100% avoidable and that we should strive toward this goal. The flaw in this idea is described in the old saying, "He who never makes a mistake never makes anything." Mistakes should be rare, but we don't want to be so afraid of making one that we end up not making much at all. Furthermore, since it's a matter of trade-offs, if someone tells me a product is perfect it may well be more expensive than it needs to be for my purposes.

When a prophet predicts the end of the world by next May 1, and May 2 shows up on schedule, the prophet tends to go into hiding. Similarly, if someone promises us that his product will have no defects, then when a defect does show up there is a tendency for him to try to cover it up. (A school principal who insists that all student failures are due to bad teaching soon finds that failing grades have disappeared. If she thinks that this is evidence that the teaching has improved, she is easily fooled.)

Those who are lukewarm about "Zero Defects" as a program sometimes say that at least it is a good goal, or that it makes a good slogan. The above arguments suggest that quality control may be better off without it, and that a better slogan would be, "Help stamp out ZD."

39
Scaling Up and Down

When postal rates increase, the Postal Service cites increased volume of mail as one reason, in addition to inflation. The average postal patron wonders why increased volume has such an effect: if a post office gets ten times as much mail, doesn't it get ten times as much income from which it can hire ten times as many people to do the bigger job without a rate increase? People think business "scales up," but it doesn't.

As we go through school we are exposed to ideas of proportionality and scale, so that we can understand maps and such. We learn to recognize the shape of California, whether it is on a small or a large map. Schools sometimes neglect the other side of the story, which is that most things do not bear proportional relationships to each other, and we should not go around assuming that they do.

When a post office, or any business, becomes ten times as large, whole new categories of expense can come into being. A new building may be needed, or a personnel department where none was needed before. New levels of supervisory personnel appear. Some of these things, of course, are mere bumps on the curve of progress, but there are more basic reasons for businesses not scaling up. If one mail sorter can effectively sort mail into some maximum configuration of slots, what happens when the mail volume goes up tenfold? Increasing the number or size of the slots and replacing one sorter with ten may result in an unworkable traffic jam. It may be necessary to add a preliminary level of sorting, or to install an automatic sorting procedure. All such changes prevent the situation from remaining the same except for scale. It's like solving a large jigsaw puzzle, where you have to devise ways of breaking the problem down into manageable parts. If you can solve a two-piece jigsaw puzzle in two seconds it doesn't follow that you can solve a 3000-piece puzzle in 3000 seconds (fifty minutes).

A famous essay, "On Being the Right Size," by British biologist J. B. S. Haldane,* points out most entertainingly how impossible it would be to scale animals up or down from their present sizes. As he says, a man 60 feet tall, in the same proportions as a man 6 feet tall, and made of the same materials, would be unable to support his own weight. Although the essay stresses sizes of animals, it mentions briefly that size is also crucial in organizations. Probably not enough attention has been given to this aspect of the structure of our society.

Observers of the business scene are accustomed to reading about how this or that company has undergone a reorganization. Sometimes the change is made for purposes of decentralization, "to enable the various divisions of the corporation to achieve their separate goals in the most efficient and profitable way." In other cases the change glorifies consolidation, "which will permit the corporation to profit from improved cooperation and communications among divisions toward reaching their common goal." Those who favor decentralization are trying to reap the benefits of smallness, while the centralizers and consolidators are looking for the benefits of large size. All of this reorganizational activity is a crude way of trying to reach an optimal position where the trade-offs of big vs. little are as profitable to the corporation as possible. If corporations remained qualitively the same as they grew, there would be no need for such activity.

*J. B. S. Haldane, "On Being the Right Size," reprinted in *The World of Mathematics,* Vol. 2, ed. J. R. Newman (New York: Simon & Schuster, 1956), p. 950.

40
More Is Less

A rich and fashionable woman is reputed to have said, "You can't be too rich or too thin." The first half of this pronouncement may be debatable, but the second half is wrong, as has been shown by a number of young people who have overdieted and died of anorexia nervosa. Anyone who actually believes the saying is indulging in linear thinking, which is a quick way to get into trouble.

One way of determining whether or not certain foods are carcinogenic consists of feeding enormous amounts to small animals and observing the results. If an ingredient of a soft drink is suspect, an experimenter feeds the equivalent of 100 drinks a day to a number of white mice. Since this is a free country, he or she is at liberty to assume that if 100 drinks a day cause cancer in 1 out of 100 mice, then 1 drink a day will cause cancer in 1 out of 10,000 mice. This would be linear thinking, and we don't know if it is right or wrong in matters of this sort. Our lawmakers endorse linear thinking by passing laws that a food additive showing such an experimental result cannot be used.

At this point it is appropriate to point out that salt has been shown to be a cause of hypertension, or high blood pressure. Too much salt and your systolic and diastolic readings will probably soar dangerously. Still larger amounts would kill you directly. Linear thinking would conclude that we should all go off salt entirely and eliminate it altogether from all of our food. We happen to know that this, too, would cause us to die. Nobody promised us that life would be simple.

Linear thinking means believing that if an excess of A causes an effect B, then twice as much excess of A will cause twice as much of the effect B, and so on proportionately. If you like more precise definitions, see your old high school algebra book. If you prefer examples, here are two more:

> If a ton of fertilizer on a field increases the yield by 20%, then two tons will increase the yield by 40%, right? Not necessarily. This linear thinking can lead to calamity, since two tons may be excessive and actually reduce the yield.
>
> If one pie in the face brings a laugh, will 20 pies in 20 faces bring 20 laughs? Not unless the audience consists of 6-year-olds or their mental equivalent.

Linear thinking works when it works, but our world is mostly nonlinear. Often less is more and more is less.

41
Large Samples and Bad News

The best golfer can't guarantee to shoot a 68 on a tough par-72 course. But if twenty of the best pros play the course, it's a good bet that one of them will do 68 or better. And if the same twenty play the course a dozen or so times apiece, somebody will shoot a 64.

A high school with 100 students may have a 6 foot 10 inch basketball superstar, but he is much more likely to be found in a school of 5000 students, and this was true even before high schools started emulating colleges by scouting for players. It's a matter of numbers, or statistics, if you will.

These are examples of the large-sample effect that says that the extremes of a large population or sample are usually more extreme than those of a small one. Statisticians call such extremes the "tails" of a distribution. People with IQs over 140 or under 60 are in the tails of the distribution of IQs. If you take samples from a population, large samples will usually reach farther into the tails than small samples will.

This effect of reaching far into the tails is one that we don't always remember is taking place. For example, if we could arrange criminals in respect to the heinousness of their crimes, we know we would expect to find the greatest extremes in the largest populations, even if the largeness of the population did not have any other effect—just from the properties of numbers. So, if we could say that this criminal is of a magnitude 5, while that one is of magnitude 10, we would find most small towns having no criminals of magnitude worse than 2 or 3, while large cities would produce magnitude 9 or 10 criminals from time to time, even without considering sociological effects.

The same goes for bad news. A national news service can usually find worse news to report than a small-town reporter can. So as news services cover more and more territory, the news becomes more and more depressing. A ray of hope here is that daily news coverage now encompasses the whole planet, so we may be in for a period in which the extremes of bad news, at least, won't become any worse.

42
Rare Floods, Rare Scientists, and Frustrated Authors

Flood control engineers speak of 20-year floods, 50-year floods, and 100-year floods. By a 20-year flood they mean the worst flood that can be expected during a given 20-year period. Similarly, a 100-year flood is the worst flood to be expected in a 100-year period. Generally a 100-year flood is worse than a 20-year flood. The average flood in a short period of time is likely to be about the same size as the average flood during a longer period, but this is not true of the worst flood.

This is the same effect that was being discussed when we pointed out that the tallest boys in a large school are generally taller than the tallest boys in a small school, even though the average height of boys is the same in each school.

This same phenomenon occurs in many places where it isn't so easily recognized. A small country of 1 million people, even after allowing for its inability to provide expensive equipment, may feel that it should make strides in scientific research comparable to those of a neighboring country of 100 million people. People may ask, "Are not our scientists as smart as those of our neighbor?" The answer is that they probably are as smart on the average, but world-changing scientific advances are usually made, not by average scientists, but by the Einsteins, Pasteurs, Darwins, or Curies. The larger country may produce ten of these superscientists per century, while the smaller country produces only one per millenium, and he or she would have no one to talk to.

Consider authors for another example. The same principle applies here, with the very best writers in a large country usually (important word) being better than any currently writing in a small country. Here there is an additional effect of population size that seems to have escaped attention:

As a population grows, the number of people with the talent and desire to be authors grows proportionally, but does the need for books grow at the same rate? It will not, except for the fact that certain people with unusual tastes may become numerous enough to demand books to satisfy them. The number of different books of fiction sufficient for a population of 1 million will be almost sufficient for a population of 100 million. This means that when the population grows beyond a certain point there may be an ever-increasing number of frustrated authors around, not to mention publishers, critics, and others in the business.

It has often been observed that specialty magazines have been replacing the general magazines that used to be so successful. Various valid reasons have been advanced to explain this, but the effect of population size per se has been neglected. A small population can support a few magazines of general interest. Such magazines might carry, for example, occasional articles of interest to various kinds of collectors. As the population grows it becomes large enough to support a magazine devoted entirely to collectors. Further growth results in enough subscribers to support separate magazines for stamp collectors, coin collectors, and collectors of rare china. As more and more specialty magazines appear, the reader (whose income does not go up with population size) has to limit his or her subscription list; if a few specialty magazines cover most of the reader's interests, the general magazine may have to go.

43

The World Is Getting Smaller

"The world is getting smaller" has become a cliché because it so pithily describes a rather complex situation. The very pithiness of the phrase may obscure some of the ramifications of the idea, so let's look at it from the point of view of some of its statistical aspects.

What is getting smaller about the world, of course, is not its geographical size or its population, but the time it takes us to move around in it or to communicate in it. What this means is that we become personally involved in a much larger group of people than used to be possible.

The good news is that we are able to see and hear finer musicians, artists, writers, athletes, or whatever entertains or interests us. The bad news is that if we *are* a musician, artist, writer, or athlete we may never acquire an audience. Until fairly recently, as human history is measured, a person might be the best singer for miles around, and many listeners might in their lifetimes never hear a better one. That was not good for the audiences, but it was good for the singer. Applied over all activities, this meant that a person who wanted to excel could usually find something to excel at, something at which he or she could become "the best in town." Small-town or small-city people who were overly impressed by such a distinction appeared provincial to their neighbors from larger urban centers, but nevertheless they could attain a degree of personal satisfaction that only a much smaller percentage of people can achieve today. A father and mother could encourage their child to become the best debater in the local school. Today that child's children or grandchildren are disappointed and may consider themselves losers if they don't win state or national championships. Psychologists can no doubt tell us what it does to our psyches when, no matter how good we get to be at doing something, nearly all of us can flip the switch on a television set and see someone do it better.

There is another kind of bad news associated with communication in our "small world," and this has to do with "feedback." To explain what an engineer calls feedback I will digress to a story about an engineering school library. Early control engineers thought it would be impressive to have a device that would turn on the library reading lights as soon as the sun went behind a cloud, and then turn them off again when the sun reappeared. They installed a sensor to detect the amount of light falling on the library's users and connected it to the electric lights. The sun went behind a cloud, this information was "fed back" to the brain, and the lights came on, a triumph for modern engineering. But then the sensor noted that now there was light and immediately ordered the electricity to be turned off; darkness resulted again, and the lights came back on, and so on ad infinitum, sending the engineers back to the drawing board. This kind of uncontrolled oscillation, known as "hunting," often happens to feedback control systems.

This oscillation can also happen to us because our communications are too good. A few years ago a national surplus of engineers developed. Such a surplus tends to be a relatively small thing, of the order of 10% or less, and it could be remedied by a small reduction in the number of students going into engineering. In a time when communications were not so swift, only a small number of students would have heard about the surplus, and things might have worked themselves out smoothly. However, we now have a vast network of counselors communicating to high school students the latest news in the employment world, and so the recent surplus in engineers became known at once all over the country. Immediately there was a sharp decrease in engineering enrollments, resulting in an overcorrection that led to a national shortage in a little over one college generation. News of this will probably lead to another surplus, and the future supply of engineers may go on and off like the lights in the engineering library.

44

Population Size and
the Quality of Life

Nowhere has the effect of population size been more dramatic than along the American highway. Two generations ago most roads were poor, service stations were few and their facilities were inferior, and it was hard to find a decent, simple meal or a good place to spend the night. A generation later motels had sprung up, as had great chains of service stations with clean rest rooms, and a vast number of spots where a hurried traveler could get a quick meal that, while it might fall far short of gourmet quality, at least wasn't poisonous. Most of this was the effect of population growth. The number of cars on the road had simply increased to the point where all these amenities had become economically feasible.

Still another generation passed, and further growth took place, but steady improvement did not occur. Some travelers are tempted, upon seeing the filth that now exists around rest rooms and some other facilities, to believe that the Age of the Slob is finally upon us. It may be, however, that they are simply observing another population-size phenomenon that did not scale up. The density of people in some areas is now greater than we know how to handle. Our system may need to be radically changed. In the 1920s most travelers had to stay at hotels in small towns. Then the increase in population allowed and encouraged the invention of the motel and of the fast-food place. These made a qualitative change in the traveler's lifestyle. We may be approaching the point where another invention is needed. On the other hand, it may fail to appear.

How to Tell the Liars from the Statisticians

Environmentalists don't always give population size as much consideration as they should. A movie on the importance of wilderness preservation shows how a couple of lone hikers can thread their way through the wilderness and carefully avoid doing any harm to the environment. They build a small fire at night from dead wood and then when they break camp in the morning they very carefully carry away any mess they made. Now imagine 100 thousand campers in the same area. There is no way that all these people could come there and interact and remain as careful with the environment as the single aforementioned pair. But suppose they could be equally careful. What would be the effect on the wilderness of 100 thousand small campfires, of the human waste of 100 thousand people? It is not sufficient to say that we will not permit many people ever to go into the wilderness area. This prohibition will be possible only if the total population stops growing. If the population continues to grow and grow, the time will come when pressure to use the wilderness for other purposes will automatically become greater than any passion for preservation, however great, will be able to resist.

Insofar as quality of life is concerned, there is a best population size for any given area. We may not agree on what it is, but it is naive linear thinking to think that science will always find ways to prevent unlimited population growth from ultimately dehumanizing society.

45
Raisins, Nuts, and Samples

When cookie batter is prepared, it has to be mixed thoroughly so that each cookie will get its share of the raisins and other ingredients. To the statistician, the batter is a population, and each cookie is a sample from it. Only if the batter is well mixed can a randomly chosen cookie be called a random sample. Extension of the sampling idea to human populations is common, important, and often badly described by reporters, so that to avoid being a victim of data pushers everyone should know something about the simple aspects of sampling.

If I put 40 pounds of raisins and 60 pounds of nuts into a barrel and mix thoroughly, then a scoopful of the mixture (a sample) will contain roughly 40% raisins and 60% nuts, by weight. If this were not true I would not consider the mixing to have been completed. If it is true, then an inspector who wants to measure the proportions of the two ingredients can take a scoopful and weigh the nuts and raisins separately; this will provide the approximate answer without having to separate all the nuts and raisins from the whole barrel.

A population of people, say, 40% Republicans and 60% Democrats, represents the same situation except for one important feature — we can't put them into a barrel and mix them up. However, if we want to estimate the proportion of voters from each party, we can assign each person a number and then put the numbers into a hopper and mix them. From this bunch of numbers, we can extract a scoopful, called a random sample if the numbers have been well mixed. Then if we contact the people whose numbers showed up in the scoop, these people constitute a random sample just as if we had mixed them up in a barrel. Finally, we ask this sample of people for their party affiliation, and we will, thanks to the mixing, come pretty close to the 40–60 split that holds for the population.

"Pretty close" reminds us that the method is not 100% accurate. Obviously you can't mix the batter so that every cookie will have exactly the same number of raisins. The variability from one cookie to another is called "sampling error," and there are mathematical ways of measuring the amount of sampling error that can be expected in a given situation.

Most people don't measure their own errors, and statisticians sometimes get the short end of the stick because they are able and willing to do so. Once I was involved in a study to find how the workers in an office divided up their time. We proposed doing a sample study by coming in at random moments over a long period of time and recording what each person was doing at each of those moments. We made the mistake of showing that the sampling error would be rather small. A nonstatistician's proposal was to stay in the office a week and record everything that everybody did. This proposal won out because "it didn't have any sampling error." Of course, the choice of a week was a sampling choice, and the difference between that week and other weeks was a sampling error. Furthermore, if there were any seasonal trends in the work, as there usually are in an office, the result would be biased according to which season was chosen for the week of study. Still this method was chosen because it was billed as a "100% sample."

All of us learn by experience. Except for pure deductive processes, everything we learn is from someone's experience. All experience is a sample from an immense range of possible experience that no one individual can ever take in. It behooves us to know what parts of the information we get from samples can be trusted and what cannot.

46
Political Polls

When political polls first hit the scene, a lot of folk mythology grew up around them, spawned by writers who often knew little or nothing about the process. Some writers felt threatened by them, in particular those whose job it was to analyze the voting situation through their own experience.

Early on the pollsters made some glaring mistakes. In 1936 the *Literary Digest* conducted a poll, then often referred to as a "straw vote," the results of which showed that Alf Landon was going to defeat Franklin D. Roosevelt. This was so far off that it helped lead to the demise of the magazine, and helped to undermine people's faith in polls, such as it was, although that particular poll used no scientific principles at all.

The basic flaw in the *Literary Digest* poll was that no attempt was made to arrange for all of the population to have about the same chance of being polled. Ballots were made available to their subscribers and to people who were on other easily available lists, such as telephone directories. In that depression year such a procedure had a strong economic bias — large numbers of people who did not have the money to subscribe to magazines or even to have a telephone were not reached by the poll. The method had worked in previous elections, but the 1936 election split the country more than usual along economic lines, and the poll did not reach a proportional share of Roosevelt supporters.

For many years it was a part of the conventional wisdom that "you can't find out about 200 million people by contacting a few thousand." In recent years election-night predictions have done so well that this isn't said very much any more. Actually, a random sample of a few thousand voters will give a very good estimate as a rule, and contrary to most people's intuition, such a sample from a nation of 100 million voters will be just as good as if it came from a city of 20,000 voters. (Going back to raisins and nuts — if the raisins and nuts are well mixed, the precision of the sample depends on the size of the scoop, not the size of the barrel. If the raisins and nuts aren't well mixed, no sample is dependable. The *Literary Digest* sample included millions of voters, but they were in no way randomly selected.)

Modern pollsters can take a survey and tell what are the limits on their error, saying for example that the Republican vote will be 55% and the Democratic vote 45%, with an error of no more than 2% on either one. So why is it that they sometimes fail? There are a lot of possible reasons.

First, sampling error always exists. If 99% of the time the error is under 2%, this means that in one case out of a hundred the error will be greater than that. If the election is close enough, the wrong outcome may be predicted. Second, true random samples are not really possible in practice

— it isn't feasible to assign a number to every voter in America and mix the numbers in a hopper. Pollsters have developed many clever ways of making their sampling procedures more and more like random sampling, and also for making the samples more and more nearly representative of the population, but the procedures are not infallible. Third, people change their minds between the time they are interviewed and when they vote. These effects, and probably others that could be listed, were present when the polls incorrectly declared Thomas Dewey the winner in 1948. Election-night television watchers can testify, however, that last-minute predictions, at least, have become very, very good.

47
Opinion Surveys

In pre-survey days, Henry Ford is reputed to have said, "The public can have any color car it wants, so long as it wants black." Now the public can roar back an answer to such statements and policies. Surveys have gone beyond political questions to cover all of our wishes and preferences, from whether we'd like to have screw caps on bottles to what we think of instant coffee or capital punishment.

Opinion surveys suffered an almost fatal setback very early when a firm asked women if they would like to have a substance with which to paint their fingernails. Loud and clear came the opinion — cheap and vulgar — but those in the beauty industry, undaunted, went about their successful task of changing women's minds. In the process they seemed to be showing how wrong a survey can be, but actually they were demonstrating that a survey result can become, or be made to become, obsolete.

Originally it was hard to sell industry on the idea of consumer surveys, since it was not easy to demonstrate the validity of results. However, the visible success of political polls in predicting election outcomes from scanty early returns has been a powerful factor in getting the public to accept surveys in general. Today if we didn't have surveys modern consumers would probably demand them in order to have their wants considered. The survey industry likes to point out that even a poor survey is a better basis for decision making than having a wealthy industrialist try to guess

101

what people want. In this vein they love to recount their story that Harvey Firestone, when told that his to-be-famous Firestone Hour was to get a Sunday afternoon slot on the radio, replied, "But that's when everyone's out watching polo!"

Now that opinion surveys have become widely accepted in business it may be time to start looking at them critically again. Starting from the premise that they're better than nothing, we can nevertheless still identify problems that show there is room for improvement.

The biggest problem, not surprisingly, is money. Given a few thousand dollars to find out if people would like to see square thimbles put on the market, a survey group is hardly in a position to take a nationwide survey. So they may restrict their activity to the small university town in which, for prestige reasons, they have located themselves. Results obtained are then valid for small university towns, but may or may not be valid elsewhere. Sometimes analysts can be reasonably sure because of the nature of a question that its answer will be much the same in university towns and

steel mill towns; but this kind of thinking, carried to extremes, can lead analysts to believe that their great experience permits them to know what people want and that they need only to do a very skimpy job to verify it.

Another problem is the loaded qustion. On highly controversial subjects it's almost impossible to get people to agree on what is or is not a loaded question. Although most consumer surveys are not exactly saturated with controversy, the possibility of accidental loading is nevertheless present, and effort is required to prevent it. Loading can also occur through a tendency to avoid direct questions for obscure psychological reasons. Instead of asking, "Do you like powdered milk?," surveys often ask, "Do you think people like powdered milk?" This second form, instead of helping us learn what people want, only helps us learn what they think other people want. This piece of information is usually of little value, so the question is loaded to lead to an irrelevant conclusion.

Finally there is the problem of personnel, which is partly but not entirely a problem of money. A good survey requires the services of good statisticians, good psychologists, and good interviewers, to mention a few. Good statisticians and good psychologists are not cheap, and furthermore they are not easily identified by persons who are not specialists in those fields. Interviewers have to be numerous as a rule, so it often isn't economically feasible to give them as much screening and training as they should have. Some research outfits have been able to survive in spite of numerous personnel deficiencies because some businesspeople, once sold on the idea of conducting a survey, will underestimate the importance of finding a qualified group to carry it out.

The usefulness of opinion surveys is no longer in question, but it still pays to view their results with intelligent and educated skepticism.

48
Television Ratings — More of the Same

No sooner had sponsors learned that they could find out how many people were watching their shows than they began asking questions like, "How many businesswomen between ages 20 and 30 are in the audience?" Some sponsors had barely begun to believe the ratings when others started canceling shows with 21.1% ratings because a competitor had 21.2%.

Working out a television rating for a particular program poses pretty much the same problems as do opinion surveys and political polls. Once again, a reasonably large random sample will provide a pretty good estimate of the number of viewers, and this is a valuable piece of information for advertisers who are plunking down the small fortunes required to buy a few moments of time on network television.

However — every geographical area is different from every other one. Surveying Atlanta will not necessarily provide useful information about Peoria. A nationwide rating requires a nationwide survey, which can be expensive and difficult. In fact, all populations are different, and the population of businesswomen between 20 and 30 is obviously different from the population of teenage boys or of billionaire hermits. Every time you try to break down the viewers according to another set of criteria, you are essentially reducing the size of the sample from which you are trying to squeeze information. This process can't go on indefinitely without either spending more money or reaching more erroneous conclusions.

Furthermore—random samples are hard to get in television, just as they are in other cases. Clever devices have to be used to produce reasonably representative samples within the limits prescribed by the sponsor's willingness to pay. Sometimes clever devices can be carried too far, so that in hindsight they become unwarranted shortcuts. As in any business, pressure to produce results that are more precise than is feasible for less money than is truly needed tempts some survey outfits to promise more than they can deliver. This is especially true in areas like TV ratings where it is not always possible for the customer to check the correctness of what has been delivered.

The hardships involved in getting a random sample have led one outfit to establish a permanent sample of viewers. This saves money, of course, but it has the seldom-mentioned disadvantage of making permanent whatever bias it has. So, if you belong to a group of viewers who happen to be underrepresented in its sample, the shows that your group would like to see will be permanently undershown.

49
The Significance of Significance

Alert readers may have noticed in recent years a great increase in the use of the word "significant." "Scientists have discovered that the weather is significantly warmer (colder) than it was a hundred years ago." Usually this word is being used in a technical sense introduced by statisticians, and it doesn't always mean what a nonstatistician might expect it to mean.

Experiments to test a headache remedy are complicated by the fact that some headaches go away by themselves, while others resist every known remedy. If ten people with headaches are given the remedy and ten others are given placebos (sugar-coated pills of no medical value), some in each group will recover in a given time and some in each group will not. Even if the remedy had no effect at all, the treated group (those getting the remedy) would do better half the time than the control group (those getting placebos). So it's not enough to observe that the treated group fares better — we have to show that their advantage over the control group is greater than could reasonably be expected to occur by chance alone. When the result is positive, that is, when the treated group does in fact do better than could be expected by chance if the remedy were truly worthless, then we call the result "statistically significant."

Since statisticians use the phrase a great deal they tend to use its shortened form, leaving off the modifier "statistically." This is fine so long as people understand, but misinterpretation is possible, in fact rampant.

How to Tell the Liars from the Statisticians

Suppose I have a headache remedy that can relieve one headache out of twenty. If I test it by comparing two groups of twenty people I have almost no chance of getting a statistically significant result. However, if I test it on two very large groups, say, of 1000 people each, I will detect this small advantage because around 50 of the headache sufferers will find relief from the remedy. Even if two or three hundred of each group get better accidentally, these fifty will stand out as an advantage for the remedy. It is then legal to advertise, "Our remedy was found to provide relief in a significant number of cases." This sounds good, but how many dollars would you pay for a small bottle of pills that help only once in twenty times?

The idea of statistical significance is valuable because it often keeps us from announcing results that later turn out to be nonresults. A significant result tells us that enough cases were observed to provide reasonable assurance of a real effect. It does not necessarily mean, though, that the effect is big enough to be important.

50
Garbage In — Garbage Out

Once during a visit to a soft drink bottling plant I was impressed by the sight of a bottle inspector who spent every boring hour of every workday watching empty bottles go by. The bottles had been cleaned by automatic equipment, but there was then no machine other than the human inspector for detecting foreign objects that might have escaped the cleaning process. I want to use this bottle inspector to make a point about the way people sometimes misinterpret statistical tests and quality control procedures.

The first part of the point is that most screening procedures are imperfect — the bottle inspector, no matter how alert he or she may be, will not reject 100% of all the bottles that should be rejected. Similarly, a significance test will not throw out 100% of the experimental results that should be thrown out.

The second and main part of the point is that when a screening procedure does fall short of perfection, the quality of the product cannot be guaranteed by the screening procedure alone. If a bottle inspector is able to detect 99% of the foreign objects in passing bottles, some people seem to think that this means that 99% of the bottles getting into the hands of consumers will be free of objects. To see that the inspector's error rate and the percent of defectives in the final output are only weakly related we have only to look at the two extreme cases. ("Defective" is a word often used to describe any item or unit that should be rejected by the screening process.)

At one extreme we have a product that contains no defectives before it gets to the inspector. Obviously, it will still have none after passing by the inspector, even if the inspector is incompetent. At the other more interesting extreme, consider a product that is completely defective. If the inspector is perfect, nothing will get by and there will be no output. If he or she is merely almost perfect, a few will get by, but they will all be defective, so that even though the error rate may be very small, the output is 100% bad. For those whose mathematical limitations prevent them from grasping this thought, someone invented the picturesque phrase "garbage in—garbage out" which describes the situation very nicely.

Many statistical tests, such as the significance test that has already been discussed, are screening procedures exactly like the bottle inspector. They detect some useless headache remedies that might otherwise have been accepted, by showing that their results are no better than might be expected from chance spontaneous recoveries. But the very fact that chance is involved means that the procedures are not perfect and some false positives will result. "Garbage in—garbage out" then applies, so that if we try a string of totally useless remedies we will eventually get a false positive. We were told as children that Edison made a successful electric lamp by trying hundreds of materials for filaments until one finally succeeded. If we did the same for headache remedies, trying chewing gum, potatoes, chicken soup, etc., in succession, we could expect to achieve eventual "success," but it would very likely be a false one.

How do we recognize the "garbage in—garbage out" situation if it can occur when the inspector is near-perfect? The telltale clue is the rejection pile. If nearly all of the bottles approaching the inspector contained foreign objects, then he or she would reject most of them, and would soon be surrounded by a great mass of rejected bottles.

In scientific investigations it isn't common to report failures, so what corresponds to the inspector's mass of rejected bottles may be swept under the scientific rug. So if we hear of some unlikely result of experimentation, such as an announcement that marshmallows cure headaches, we should ask, before assigning credibility, "Could this easily have been the end result of a series of unmentioned tests in which many useless treatments were tried?"

As usual all this is greatly oversimplified to make a point. Science usually amounts to a lot more than blind trial and error. Good statistics consists of much more than just significance tests; there are more sophisticated tools available for the analysis of results, such as confidence statements, multiple comparisons, and Bayesian analysis, to drop a few names. However, not all scientists are good statisticians, or want to be, and not all people who are called scientists by the media deserve to be so described.

Thoughtful observers of the significance test procedure will point out that if we make it harder to pass the test, we won't get so many false positives and it will become harder to obtain false results through sheer persistence. Unfortunately, this would also have the effect of rejecting many of the beneficial treatments along the way, until a treatment could be found that would work 100% of the time and so would be always accepted. It comes down to the question, "Why can't we go back to the good old days of Tom Edison?" Edison had it easy—it doesn't take statistics to see that a light has come on. When we find a cure for headaches or cancer that works as consistently and as fast as we can turn on a light, we won't need any statistics to confirm that, either.

51
The Double Negative and the Consumer

Does this claim sound familiar?: "Independent laboratory tests show that no other leading product is more effective than ours." Translation: "A small test was run among the leading products, and no significant difference was observed among the products tested."

A coin that comes up heads 60% of the time would be called "loaded," but its dishonesty could not be detected in 10 throws. If we toss it 10 times and get 6 or 7 heads, we would have to admit that a perfectly balanced coin might do the same. However, if we toss the coin 1000 times and get 600 heads, we have a significant result because we believe it would almost certainly not happen by chance to a perfect coin.

In general a small-scale test or experiment will not detect a small effect, or small differences among various products. This is bad news to most of us since it means that good tests are expensive, but there are people around who can make good news for themselves out of almost anything. Suppose we want our product to come out of a test with a favorable result compared with a better product. We simply run a very small experiment that will detect only very large differences. When no "significant" difference is observed, we say something like, "No evidence has been discovered to show that our product is not better than theirs." Obviously the reader of this double negative statement is not supposed to notice that it doesn't say anything. As a cheerful attempt to make nothing look like something, such statements compare favorably with that great advertising classic, "Prices reduced up to 30% or more."

111

52
The Double Negative
and Social Science

It is often said that "there is no evidence that punishment is a deterrent to crime." When this statement is made we are apparently supposed to conclude that punishment is not a deterrent to crime. This may leave us more than a little confused as we wonder why those long lines of vehicles form behind the patrol cars on our highways, or why looters so often show up just when the police are out on strike.

Whether punishment is or is not a deterrent to crime is a question that pertains to the complex human behavior of all four billion people on earth. It depends not only on the people themselves, but also on the swiftness, certainty, and severity of the punishment.

The experimental method, which has been so successful in the natural sciences, has been creeping slowly into the social sciences. Where it is applicable it has enabled the social sciences to make progress, but it doesn't apply as often as we might like. Realistic experiments with people as subjects are not always possible, and just looking at data can often be misleading, as we've already described. Social scientists and psychologists sometimes try to overcome these handicaps by conducting small experiments whose subjects are college students; the latter may be looked upon with disfavor by their professors if they don't consent to be guinea pigs. Small-scale unrealistic experiments on atypical involuntary subjects often lead to unwarranted conclusions.

In questioning whether punishment is a deterrent to crime, we have to remember that the behavior of individuals is highly variable, and that the answer is highly dependent (as mentioned before) on whether the punishment is sure, on whether or not it occurs quickly, and on how severe it is. It is probably not possible to measure the influence of all these factors in an experiment, so if the experiments we can run fail to show any deterring effect, we should avoid jumping to general conclusions. A good experiment is better than an authoritarian judgment, but this doesn't mean that a poor experiment is more valid than our logic and common sense.

53
What Is Correlation?

One of the blessings conferred on society by statisticians is a numerical way of measuring correlation, such as is alleged to exist for example between redness of hair and fieriness of temper. This measure, called the correlation coefficient, is a useful tool when properly used, but so are razor blades and guns.

Since "correlation" is a term used in a general sense by laymen, I thought it might help to go to a dictionary for a layman's definition to use as a starting point. Various definitions are available, one of the better ones being, " . . . Reciprocal or mutual relation in the occurrence (as of deafness in blue-eyed white cats or the expression of apical dominance in plants) of different structures, characteristics, or processes in organisms. . . ."* Now if this is exactly what you need to make everything clear, feel free to skip the next paragraph.

We can show that people's weights are positively correlated with their heights. By this we don't mean that every woman weighs more than all women who are shorter than she is, but that, *on the average* weight increases with height. The correlation coefficient is the measure of the

*Reprinted with permission from *Webster's Third New International Dictionary.* Copyright 1981 by G. & C. Merriam Company, publishers of the Merriam-Webster dictionaries.

strength of this relationship. Perfect positive correlation is measured at + 1, and would occur if weight above (or below) average were exactly proportional to height above (or below) average. Less than perfect positive correlations are measured by numbers between 0 and 1, with 0 indicating no correlation at all. (Strictly speaking, a 0 correlation coefficient means only no linear correlation. I mention this only for the statisticians in the audience, but surely they have already found that we are not speaking very strictly here.) Negative numbers are used for negative, or inverse, correlations exemplified by weight and running speed — some heavy people can run very fast, but in general speed goes down when weight goes up.

As soon as the formula for calculating the correlation coefficient became generally known a field day was had by all. Everything was examined in relation to everything else for possible correlation, and many were the M.A. and Ph.D. degrees awarded therefor. In spite of the "beautiful but dumb" stereotype, beauty was found to be positively correlated with brains. Teachers' salaries were found to be positively correlated with the sale of alcoholic beverages. Lung cancer was found to be positively correlated with smoking. Some of these conclusions were illuminating, some were irrelevant, many were accidental, some were incorrect, and some were misleading.

The most common mistake in interpreting correlations is to suppose that they demonstrate a cause-and-effect relationship. In our observations on smoking and lung cancer it was pointed out that the fact that they are positively correlated, while it suggests what to look for, does not in itself demonstrate that smoking causes lung cancer. I won't add to the voluminous discussions that have already appeared on this aspect of the interpretation of correlations, but there are various other sources of misinterpretation, one of which will be discussed next.

54

The Coin that Won't Stand
on Edge

Correlations make good news stories. If a correlation is found between financial condition and marital breakups, a story for the Sunday features will result. A correlation of grades in school with later job success, especially if it is negative and so makes a lot of people happy, will rate an article on page 2 or 3. A page 1 write-up is assured if anything at all is shown to be correlated with cancer, heart disease, or sexual activity.

The beauty of this from the data pusher's point of view is that it matters little how the result comes out — that is, whether the correlation is positive or negative. Only a 0 correlation is uninteresting, and in practice 0 correlations do not occur. When you stuff a bunch of numbers into the correlation formula, the chance of getting exactly 0, even if no correlation is truly present, is about the same as the chance of a tossed coin ending up on edge instead of heads or tails.

If there is truly no relationship between sunspots and terrorist bombings, for example, some data collectors will find small positive correlations and others will find small negative ones, the group as a whole averaging very nearly 0. So this is a place where the concept of "statistical significance" should be used. If the observed result is no farther away from 0 than might be expected to occur by pure chance, we should forget the apparent result, or at least avoid becoming excited about it unless further investigation produces significance.

Suppose a publicity release does say that sunspots and terrorist bombings are correlated. If the release doesn't use the word "significant" it's a good bet the result has not passed a significance test, since "significant" is not the kind of word your average reporter is likely to omit once it has come up. If the word does appear, credibility still does not automatically result. It can always be argued that "significant" is an everyday English word of imprecise meaning, so that in a land of free speech and freedom of the press anyone is entitled to use it. Furthermore, even if the word is used in the precise sense of statistical significance, we've already shown how it does not necessarily follow that the relationship is strong enough to be of any importance.

Finally, if a correlation is both significant and strong, we may still want to inquire into its relevance. If a psychiatrist announces a correlation between stomach ulcers and early childhood repression, we don't have to believe that the conclusion applies across the board. Unless there is evidence to the contrary, it is reasonable to believe that the psychiatrist did not go out into the streets to obtain a random sample of people. Therefore, the correlation is probably relevant to the restricted group of people who show up on that particular psychiatrist's couch, and not necessarily to the population at large.

55
Guilt by Association

Young male drivers have a reputation for trying to assert their manhood through various forms of reckless driving. Such behavior can be expensive to insurance companies, who retaliate by assessing all young male drivers a higher premium. It is true that collectively we must somehow pay for the bad habits of a few, but a young male driver who drives sensibly may wonder why he has to shoulder more of the burden than other drivers do.

The insurance company's position is that accidents are positively correlated with male youthfulness, and that this justifies higher rates for those who are indicted by this correlation. One alternative would be to spread the cost evenly over all drivers. Another alternative, more popular with good drivers, is to increase premiums sharply for drivers who are deemed at fault in accidents. Actuaries say they dislike this latter course because the statistics produced by an individual are usually too few to be reliable. In other words, if a young man is at fault in an accident one year after he learns to drive, this may be the only accident he will ever have, so it doesn't necessarily predict his future; on the other hand, if the insurance company withholds action until he has had several accidents, he will in most cases no longer be young when it is finally decided that he used to be a dangerous young driver.

Instituting a sharp premium increase after just one accident may be unjust, but the question is, "Do we create more injustice this way than we do by grouping people by correlations?"

Sometimes the argument for using correlations to determine insurance rates for drivers is based on analogy with life insurance premiums. Few people need to be convinced that life insurance premiums should depend on age and state of health, even though examples abound of young insurees who died before older ones. A strong reason for basing life insurance premiums on age, however, is not so much the fact that death rates are correlated with age, but that if everyone were charged the same rates the public could take unfair advantage—octogenarians and people with terminal illnesses could bankrupt the insurance companies by taking out large policies. Spreading out the costs of young male drivers' accidents over all insurees, however, would not lead to comparable action. No young male driver would say to himself, "Hey, this insurance is a great bargain for a bad driver like me. I'll take a million dollars' worth."

I have in front of me a newspaper article that tells how an insurance company divides drivers into "clean risks" and "dirty risks" using correlations. "Dirty risks" once included divorced women, house painters, barbers, and blacks, to name a few. Some of these categories have been eliminated by public pressure and by antidiscrimination laws. Others, like the house painters and barbers, will presumably continue to be discriminated against unless they organize on a national basis. Incidentally, house painters and barbers are said to have been declared "dirty risks" because "the companies decided these occupations had a reputation for drunkenness."

Using correlations to determine insurance premiums can be a form of applying guilt by association, a practice consistently frowned upon when it occurs in other areas. If we are going to decide intelligently which correlations are "fair" and which are not, we first need to understand what correlations are and how they are arrived at. We especially need to understand that it may take years to effect removal of a category for being discriminatory, but it takes only a few minutes to suggest a number of new categories to replace it. No matter how safely you and I may drive, it's probably easy to identify categories of people that (1) include you and me, and (2) have an overall driving record that is not as good as average.

56
Psychological Tests and Job Success

Job applicants are sometimes required to answer questionnaires that cover such personal items as sex life or preferences among television shows. Some consider this an invasion of privacy. Others feel intuitively that something about these tests is wrong, but they're not sure what. One of the things that's wrong is that the tests are often based on some rather weak correlations.

It costs money to find new employees, and employers naturally look for all possible ways of making the search more efficient. An employer generally has not the slightest interest in a prospective employee's TV viewing habits, for example, unless it can be shown that these habits are correlated with success or failure on the job. If, say, employees who watch as many as five situation comedies a week are estimated to be 10% more likely to fail in their jobs than others, the employer will compute exactly how much money can be saved by never hiring any situation comedy fan.

The victims of such a policy are the individuals who would make good employees but who happen to be watchers of situation comedies. A computer has used correlation to associate them with a group of individuals who, on the average, might not do well at the job in question. They may well be justified in feeling that they have been unfairly discriminated against or that guilt by association has been applied to them. Correlation analysis is a useful tool for uncovering a tenuous relationship, but it doesn't necessarily provide any real understanding of the relationship, and it certainly doesn't provide any evidence that the relationship is one of cause and effect.

People who don't understand correlation tend to credit it with being a more fundamental approach than it is. Although methods for developing psychological tests have become fairly complex and sophisticated, their basis can be described in simple terms as follows: start with a sample of employees and grade them according to the quality of their work. This grading may be as crude as merely dividing them into two groups—satisfactory and unsatisfactory. Then prepare a very long list of questions and ask the employees to answer them. Looking at each individual question, compare the answers of the satisfactory employees with those of the unsatisfactory employees. If the two groups answer this question in pretty much the same way, the question has no discriminatory power, so eliminate it. If the two groups answer the question somewhat differently, apply a statistical test to determine whether the difference is statistically significant. The questions that pass this significance test are used to make up the job application test.

While the use of significance tests is better than nothing, the usual troubles show up. We may well have a "garbage in—garbage out" situation. That is, if all or nearly all of the questions tried are really irrelevant (i.e., will have essentially no discriminatory power when used on future groups of applicants), then the same will be true of all or most of the questions that finally appear on the test. Also, even if many of the questions do have discriminatory power, they will tend to have less than they appear to have by virtue of the "regression fallacy" discussed earlier. This means that the employer who calculates the amount of money to be saved by not hiring certain classes of people will almost certainly overestimate such savings. The true savings may not be enough to counteract the cost of generating the greater flow of applicants that are needed because the psychological test eliminates many of them.

There is no need here to go into the benefits of psychological testing of job applicants. These are amply publicized by those who make and sell the tests. The purpose here has been to point out that statistical tests and computers, wondrous things though they may be, are not able to turn a poor psychological test into a good one.

57

Credit and Your Computer Kin

Andrew Carnegie is said to have established credit by starting with a very small loan, repaying on time without ever using the money, and doing this in ever-increasing amounts until his credit was practically unlimited. This in no way proved that Carnegie was dependable, but merely that he was clever. What he was doing was beating the credit system which in those days was based largely on references, a useful method in smaller populations. The heavy use of references had the advantage that it encouraged people to value their reputations, but it also led to nepotism and the belief that "it's not what you are that counts, but who you know." Young Andrew didn't know any influential people, so he did what he could.

Our population has grown and it is no longer possible to handle credit entirely on a person-to-person basis, so part of the job of establishing credit has been turned over to computers. If you want to find out whether computers have correlated you with the "right" kind of people, just set out to borrow a few thousand dollars. Nepotism is on the wane and you are judged now not by your relatives, but by your correlatives. You may seldom meet any of your correlatives, but your computer card and theirs (or whatever is the latest equivalent in modern computer hardware and software) will be side by side in the same slot after sorting. They are the people who share your age, sex, marital status, job type, educational background, or other less relevant characteristics that someone may have decided to

record about you. This someone can take any facts about individuals, use a computer to calculate which ones seem to be correlated with good records of paying off debts, and then set up rules that determine who gets credit. The person who does all this may or may not have any understanding of the strengths and weaknesses of the statistical tests involved.

You yourself may be a good credit risk, but if your correlatives are not you will find it hard to borrow money. This is not to say that there is any obvious way of improving our present system, but it's worth remembering that life doesn't automatically become fair just because we've used a computer.

58
College Entrance Tests

Do college entrance tests really work? Some recent studies claim to show that they don't, that is, that grades on entrance tests are not well correlated with success in college. Before we examine the statistical fallacy behind these claims, a little background is in order.

Admitting unqualified students to college is a waste of their time and money, and their presence also tends to make the educational process less efficient for others. Before entrance tests came into general use, colleges had to rely on high school and prep school grades in deciding who should be admitted. Standards of grading vary so much from one school to another that colleges tended to accept students only from those schools whose standards were known to be high enough; at some prestige universities the result was that their student bodies came primarily from a small number of prep schools. Letters of recommendation had the same drawbacks as grades, and their general tendency to be favorable grew beyond all bounds when people began to be afraid of lawsuits resulting from unfavorable letters.

Entrance tests were invented to make the admissions process more democratic. Early tests were of the "essay" type in which the student was required to write sentences or paragraphs in answer to questions. It turned out that this kind of test left too much to the judgment of the person who did the scoring. In fact, at least one study showed that the scorer was more important than the student in determining the score. The "objective" or "multiple choice" test then replaced the essay test, removing this disadvantage but introducing others.

The tests most commonly used today are of two kinds: the aptitude test which estimates various aspects of intelligence and aptitude for college work, and the achievement test which attempts to measure what the student has learned. These tests do not measure desire, creativity, or other more intangible characteristics, but a good score on the achievement test generally indicates that the student has had some desire to learn in the past. The tests worked in that it was observed that people who did well on them tended to do well in college, while those who did not do well on them tended to do poorly in college. No one claims that exceptions don't exist. This is a correlation, with its usual advantages and disadvantages. If all colleges had the same admission requirements, complete dependence on the same admissions tests would be unfair to those individuals who, for one irrelevant reason or another, do not perform up to their potential on the tests. However, they are protected against this danger by the fact that our enormous variety of colleges and universities provides a place for almost everyone, regardless of their test scores. The tests can and should be made better, but they should not fall victim to the if-it-isn't-perfect-destroy-it philosophy.

So how is it that studies are now cropping up that show that the tests do not correlate with success in college? The post-World War II "baby boom" produced a glut of college students in the middle 1960's, so that most colleges could find room for only a small percentage of applicants. They were therefore able to pick and choose, filling up their campuses with students who had done unusually well on the entrance tests. Studies carried out in such schools often show little correlation between entrance test scores and later college grades. The reason is not that the tests failed in their mission of identifying superior students for admission, but that they are not designed to differentiate among the top students. The difference between a student who makes a very high grade on an entrance exam and a student who makes a very, very high grade depends on a small number of test questions and so is heavily influenced by chance. Anyone who has had experience with classes that are really heterogeneous knows that it takes only one or two questions to obtain a fairly good prediction of who is going to fail. In any correlation situation if we look at only one end of the distribution of individuals, the correlation will not look as strong as it is over the entire group. The studies that I have seen that claim to show no correlation were studies applied to student bodies from which the really unqualified students had already been weeded out.

There remains a question that no one seems to have asked: if the regular college entrance tests are not very good at distinguishing among superior students, why are they so widely used in deciding which superior students will receive scholarships of various kinds and which will not?

59
Science and Statistics

Children are told that an apple fell on Isaac Newton's head and he was led to state the law of gravity. This, of course, is pure foolishness. What Newton discovered was that any two particles in the universe attract each other with a force that is proportional to the product of their masses and inversely proportional to the square of the distance between them. This is not learned from a falling apple, but by observing quantities of data and developing a mathematical theory that can be verified by additional data. Data gathered by Galileo on falling bodies and by Johannes Kepler on motions of the planets were invaluable aids to Newton. Unfortunately, such false impressions about science are not universally outgrown like the Santa Claus myth, and some people who don't study much science go to their graves thinking that the human race took until the mid-seventeenth century to notice that objects fall.

A history book for young people will mention the development of anesthetics in just two or three lines. The impression will be given that the experiment consisted simply of giving various drugs to a volunteer until sleep occurred, at which point the traditional "Eureka!" was shouted and the experiment was complete.

It was not like this. Experiments with anesthetics have to determine how much of a drug is needed to be effective for any desired length of time, how much of a dose will result in undesirable side effects, including death, and how these amounts vary from person to person. Finding the optimum dose for each of various drugs is an elaborate data-collecting procedure. The search for new and better drugs never ends, nor does the search for better understanding of the properties of already-known drugs.

Statistical reasoning is such a fundamental part of experimental science that the study of principles of data analysis has become a vital part of the scientist's education. Furthermore, as must be obvious to readers of this book, the existence of a lot of data does not necessarily mean that any useful information is there ready to be extracted. So there grew, along with the studies of data analysis, the discipline of Design of Experiments, which has to do with the principles of efficient planning of data-collection procedures. Surprisingly, it took until well into the twentieth century for these subjects to take definite shape, with the result that there are still large areas of science in which data collection and analysis are done in amateurish ways.

60

The Scientific Method
Authoritarianism In and Out
of Science

Rulers of ancient Egypt let it be known to the populace that the sun did not rise until bidden to do so by the Pharaoh. Having control of the sun was a great help in maintaining a position of authority, and this idea was used in various early societies. On a more sophisticated level, those who learned how to predict eclipses found it profitable to claim credit both for darkening the sun and for allowing it to return, just as Mark Twain's fictional Connecticut Yankee did at King Arthur's Court.

Authoritarianism thrives on ignorance, and in early civilizations those who could achieve an advantage of knowledge often used it as a base for founding powerful priesthoods. But authoritarianism is not limited to priests. It's safe to say that in almost any group that happens to obtain exclusive use of certain information there will be individuals who will consider using it to their own advantage in developing a power base. Science is one of our best weapons against authoritarianism, but authoritarianism has been known to surface among scientists. When this happens, misguided perfectionists or romanticists sometimes seek to root it out by attacking science. Instead of destroying science, which would merely return us to ignorance and superstition, what we need to do is to expose and root out the authoritarians.

One of our best weapons in the fight to spread the truth is the scientific method. In simple form this method has always been with us, but its

beginning as a formal approach is often equated with Galileo's experiments on falling bodies. For centuries before Galileo, authoritarian philosophers had proclaimed that the speed of falling bodies depended on their weight. In an age when authorities were usually believed, people had no trouble believing this, especially since anyone could see that rocks fall faster than feathers. Galileo questioned the authorities and performed his storied experiment by dropping metal spheres of various weights from the Leaning Tower of Pisa. The fact that they all landed at the same time helped put an end to notions that had endured for centuries. The slow fall of feathers turned out to be a matter of air resistance, not weight, for in a vacuum they fall as fast as rocks.

One basic part of the scientific method is experimentation; to find out what happens when objects fall, instead of asking a "wise man" we drop some objects and observe what happens. If someone else claims to have already done this and we're not sure of his or her credibility, we try it ourselves. When surprising results are announced, other people perform experiments to verify them, and if these experiments agree with the first one, the results become a part of accepted science.

Unfortunately, the cheap experiments have mostly been done, and we are left with questions that are expensive to answer even once, so that we cannot afford to do as much verification as we would like. What then is our defense against persons who might try, consciously or unconsciously, to deceive us? General knowledge is a great help, but we can't become experts in every field. Today's scientific investigations are so complicated that even experts in related fields may not understand them well. But there is a logic in the planning of experiments and in the analysis of their results that all intelligent people can grasp, and this logic is a great help in determining when to believe what we hear and read and when to be skeptical. This logic has a great deal to do with statistics, which is why statisticians have a unique interest in the scientific method, and why some knowledge of statistics can so often be brought to bear in distinguishing good arguments from bad ones.

61

Looking Backward
vs. Looking Forward

A sportswriter once wrote a column in which he concluded that colleges having losing football teams are too quick to fire their coaches. To support his conclusion he quoted statistics that clearly showed that colleges that change coaches frequently don't win as often as do colleges that stick with one coach. The fallacy in his argument is so obvious that I hope he was writing tongue in cheek, but there was no sign of this.

We can easily see the error in the sportswriter's "logic" because we have additional knowledge not contained in his data. We know that teams first go into losing streaks and *then* fire their coaches; since the losing happens first, it can't be the effect of the firing. A man from Mars, however, faced simply with the data, might not be able to distinguish cause from effect.

Whether the sportswriter was being facetious or not, the example is a useful one because many studies reported in the papers put us into the position of the Martian—that is, we don't have any information other than the quoted statistics — and we tend to accept the conclusions that accompany the report. How can we reduce our gullibility in this respect?

One important way of developing our powers of discrimination between good and bad statistical studies is to learn about the differences between backward-looking (retrospective or historical) data and data obtained through carefully planned and controlled (forward-looking) experiments.

How to Tell the Liars from the Statisticians

The sportswriter's data were retrospective, that is, he simply collected the data that were there. If it were possible to experiment in this area, we might do something like this: we would take a selection of colleges and divide them into two groups depending on whether they are winners or losers over a period of time in which each college keeps its coach. Then we would randomly select half the winners and have them replace their coaches while letting the other half keep theirs. We would do the same with the losers. Then after another period of time we would look at the records and see whether the schools that kept their coaches played significantly better or worse than the others.

Clearly the experimental approach is better than the retrospective one, but just as clearly it is not feasible to carry out an experiment affecting people's lives to such an extent.

When a retrospective study is reported in the media, readers or viewers should look for a possible confusion of cause and effect, as in the case of the sportswriter. They should also look for improper aggregation, or confounding. These are illustrated by statistics that show that married people live longer than others. Such data are contaminated by the existence of blocks of people who have, say, medical problems that cause them to be less likely to marry and also to die earlier than normal people. If you want to know if marriage really increases longevity, you must somehow compare married people with unmarried people who had similar prospects for longevity before making the decision whether or not to marry.

Nonscientists tend to think of science as being a lump of knowledge all in one piece, and of scientists as being all the same kind of people (the all-Chinese-look-alike fallacy). Scientists differ and so do branches of science. Physics and chemistry, for example, are probably the most experimental sciences. Sociology has to be largely observational, since we can't often experiment with people in ways that would be desirable from the point of view of the logic of drawing conclusions. Astronomy is part of physics, but we can't experiment with planets and stars as we do with electric currents or lenses, so much of astronomy is observational. But all sciences are becoming more experimental as practitioners become more ingenious in designing procedures and more skilled in keeping out statistical fallacies.

Being experimental, however, doesn't necessarily make a scientific study entirely credible. One weakness of experimental work is that it can be out of touch with reality when its controls are so rigid that conclusions are valid only in the experimental situation and don't carry over into the real world. Psychology is particularly vulnerable to this because experiments in psychology are often greatly oversimplified compared to the complex human behavior to which the conclusions are supposed to be applied.

Some of these remarks are also oversimplified. They are intended merely to give the layman a few clues to look for in deciding what to believe and what not to believe.

62
Controlled Experimentation

In the 1950s an experiment was carried out to see whether coastal storms along the eastern seaboard could be modified by cloud-seeding procedures. One newspaper account reported that the experiment failed totally. Actually the experiment was a success in that it determined the truth, which was its objective. The truth was that seeding a storm from the air and ground with dry ice and silver iodide didn't produce any large effects. It just happened that this was not a very interesting truth.

Meteorologists in charge of this experiment consulted statisticians about the planning, and the statisticians recommended a controlled experiment, which was then something of a novelty in that field. Until then many cloud seeders had taken a somewhat undisciplined approach, seeding clouds and taking credit for whatever happened subsequently anywhere (after the fashion of the rooster who thought he caused the sun to rise every day.)

The statisticians proposed that the meteorologists would decide which storms to seed and when, and that in every case a plane would go into the air, ostensibly to seed them. However, from each pair of storms one, a "control," would be selected to be unseeded; the selection was to be made randomly and without the knowledge of the meteorologists as to which storms were seeded and which were not.

The meteorologists rejected this proposal, saying that storms were few, time was limited, and they could not afford to waste half of their opportunities by not seeding them. They then left to do their own experimentation. Returning a few weeks later, they reported that they had seeded a storm and that the storm turned out to be an unusually large one, but they saw that there was no way of knowing whether it would have been just as large if they hadn't seeded it, so they were ready to try the controlled experiment.

When an experiment reveals that truth can be dull and is not necessarily stranger than fiction, it can be disappointing to the experimenter. Some experimenters develop a distaste for controlled experimentation, since it may inhibit their wheeling and dealing. The rest of us, however, should feel better when we see, in the write-up of an experiment, that adequate controls were used.

63
Cause and Effect

Alert human beings, hit by something unpleasant, seek to find its cause so that they can avoid it in the future. Primitive people tried to explain various phenomena by inventing a god for each one. Superstitions were born as individuals tried to blame catastrophes on whatever events had occurred just before, as when a person might have a heart attack shortly after breaking a mirror or having his or her path crossed by a black cat.

Sophistication brought data collection, a business that paid off well in early times in the field of astronomy. Many societies became quite proficient at predicting the motions of the planets and stars based on past observations.

But mere data collection fails when it is applied to situations in which various possible causes abound and astronomical regularity is not present. Examples have already been given here, but here's another one: data showed that students from "prestige" universities did better than students from other schools when they took the Graduate Record Examination for admission to graduate school. This immediately led many to believe that a student's money would be well spent by going to one of the expensive schools. At this point experienced data observers would (and did) ask, "The prestige universities get better students to begin with, so might that not account for their better showing on the graduate examinations?" A detailed study showed pretty much just that: taking into account the scores that individual students made on their aptitude exams at the time of entering college, this study showed little or no difference on the Graduate Record Examination according to whether or not the students attended prestige universities.

Immediately another possibly false conclusion was drawn by many. They saw this last result as indicating that the prestige schools did not provide anything for their extra cost. However, nationwide uniform examinations of any kind generally consist of multiple-choice questions which tend to measure acquisition of factual information rather than the more intangible aspects of education. Prestige universities have not claimed that they present factual information much better than other schools do; the contribution that they do make is intangible and difficult to measure.

Mistakes arising from retrospective data analysis led to the idea of experimentation, and experience with experimentation led to the idea of controlled experiments and then to the proper design of experiments for efficiency and credibility. When someone is pushing a conclusion at you, it's a good idea to ask where it came from — was there an experiment, and if so, was it controlled and was it relevant?

64
Value Judgments and Planning

A group I once worked for had this motto: "If something is not worth doing, it is not worth doing well." Having this motto didn't mean that we always agreed on what was worth doing, but it did seem to insure that at the beginning of every new project someone would ask, "Is this worth doing?"

Examples have already been given to show how retrospective data can lead to false conclusions, through mixing cause and effect, blaming the effects on the wrong causes, or failing to understand probability. When experiments are possible we can obtain data that are mostly free of these pitfalls, but only if the experiments are planned with that in mind.

First and foremost an experiment should have a goal, and the goal should be something worth achieving, especially if the experimenter is working on someone else's (for example, the taxpayers') money. "Worth achieving" implies more than just beneficial; it also should mean that the experiment is the *most* beneficial thing we can think of doing. Obviously we can't predict accurately the value of an experiment (this may not even be possible after we see how it turns out), but we should feel obliged to make as intelligent a choice as we can. Such a choice is sometimes labeled a "value judgment."

On the question of what is worth achieving there are, as usual, two extreme positions. At one extreme are those who point out that the history of experimentation is dotted with successes that resulted from haphazard

or at least unpromising-looking activity. After all, who at the time would have predicted enduring fame for Galileo's falling-body experiments? Some experimenters who love their work think that such successes should persuade us to underwrite whatever investigations their curiosity leads them to perform, and they even try to discredit the whole idea of making value judgments.

At the other extreme are those who don't want to support any investigation unless it promises immediate and certain profits. This position often does lead to positive benefits, but they tend to be very minor.

So it comes down to this: we should spend some of our limited resources on experiments that might produce great benefits but that may not have a high probability of success, but we should not put all of our money on high-risk ventures because we may end up with no benefits at all.

After it has been decided that a certain goal is worth achieving, an experiment can be planned to achieve it. The desired experiment often turns out to be too expensive, so the next step is to reduce the scope of the experiment while preserving a reasonable chance of achieving the stated goal. If the smallest such experiment is still beyond the available budget the experiment should probably be dropped. Too many experiments are planned to fit a budget, with little thought given to whether or not they have a chance to reveal the desired information. This way of doing things, though inspired by a desire to be economical, is actually wasteful of money since it leads to so many useless expenditures.

Too many experimenters believe in the old line, "You get what you pay for." In experimental work it's easy, through insufficient planning, to pay a lot and get nothing.

65
Serendipity — Puttering vs. Planning

An old fairy story, "Three Princes of Serendip," featured frequent marvelous discoveries made by pure chance. Horace Walpole poked fun at this sequence of improbable events by coining the word "serendipity." Two centuries later this word seems to be still gaining popularity and may turn out to be Walpole's most enduring monument. With the single word "serendipitous" we can describe a discovery as accidental while simultaneously expressing a slight suspicion that its accidental nature is a fairy story.

More modern stories have it that many great scientific discoveries have been made by chance while the investigator was looking for something else. The resemblance of these stories to the Newton apple story suggests that some of them may be misleading. Creative people often find it hard to explain why they did what they did, and they sometimes invent simple explanations to fend off the inquisitive. Most of us could putter around forever and the result would be just so many mud pies, so it seems likely that many beneficiaries of serendipity were, like Christopher Columbus, on the track of something big.

Society should and does encourage a small number of exceptional scientists to putter as they are inclined, but to use such a shotgun approach for all science and technology would be expensive and inefficient. Although serendipity sometimes leads to revolutionary new ideas, the development of knowledge at more pedestrian levels requires planning in order to avoid waste of time and money. The first and sometimes most useful function of the statistical consultant in the design of an experiment is to ask such questions as, "Why are you doing this experiment? What kind of things are you trying to learn? What will be your actions depending on the outcome of the experiment?"

I have seen experiments terminated before they began because these questions were asked. In some cases the experimental results would call for some action, and when it was determined in advance that the action would be the same no matter how the experiment turned out, it became clear to all that the experiment was unnecessary. In other cases the experimenter, in trying to answer these questions, began to realize that he or she really hadn't thought through the problem enough, so it was back to the drawing board for a while. Far more numerous are the experiments that turned out to be useless and that could have been avoided if someone had asked these questions ahead of time.

Experimentation is such a productive activity that it has come to be regarded in some circles as an end in itself. This view is especially popular among experimenters, who experiment because that's what they like to do. In this respect the word "experimenters" refers not just to scientists but also to artists, composers, architects, writers, and the like. Often they would have us admire their work because it is "experimental" or "innovative," regardless of whether it has any other virtue. The object of experimentation is improvement, and an experiment should lead to something beyond itself.

"Monday morning quarterbacks" know how the weekend's football games could have been won, just as bridge players conducting a "postmortem" can tell you how you should have made the small slam. People who try to make their hindsight look like foresight deserve the opprobrious term "second guessers," but there is virtue in hindsight if through its use we develop our foresight. In experimental work it's easier to see how the experiment should have been planned after it has been carried out than it was in the beginning, but we can learn from our mistakes. Statisticians who do consulting in the area of experimentation become involved in a great variety of experiments, so that they can bring to the planning stage experience that helps them to foresee most of the possible pitfalls.

66
Unconscious Dishonesty

Even people with the utmost integrity are not always able to keep their conscious desires from subconsciously affecting their work and their conclusions. (For that matter, utmost integrity is not so common a commodity that we can safely assume its presence anywhere.) A researcher performing tests to find a diabetes cure may win fame and fortune if one of them is successful; otherwise it's back to the lab and obscurity. Under such conditions it's unrealistic to expect that the human beings involved will always be 100% objective in their "search for truth."

It has come to be recognized as good experimental practice to "blindfold" some of the experimenters and their subjects in certain types of experiment. For example, it's important not to let the subjects in an experiment know which of them are receiving the prospective cure and which the placebos, since psychological effects might otherwise spoil the conclusions. If possible, even the doctors are not allowed to know, since the doctor's attitude may affect the patient's evaluation of the relief provided by the treatment. When neither doctor nor patient knows who got the placebos, the experiment is called a "double blindfold" or "double blind" experiment. One famous test is even said to have been spoiled by an overenthusiastic receptionist who, knowing which patients were receiving the genuine treatment, was so encouraging in her conversations with them that they actually reported improvements that had not occurred.

How to Tell the Liars from the Statisticians

In the previously mentioned experiment for seeding storms to increase rainfall, the meteorologists decided which storms to include in the test. The decisions as to which ones would be actually seeded with dry ice were made randomly and without their knowledge; otherwise, someone who hoped for a positive result might have, consciously or not, decided to use the real seeding only on those storms where there appeared to be a good chance of success.

Experimenters sometimes regard statisticians as paranoid on this subject. The literature, however, is full of reasons for insisting on it. In one experiment to determine the effect of milk on school children's weight, a teacher gave the milk to the children who appeared to need it most, and used the healthier ones as controls. This was a humane act, but it ruined the experiment. In another famous case, a ship captain was given a seasickness remedy to try. He gave the placebos to the passengers and the real pills to the crew, on the grounds that the latter were more important to the success of the voyage. These rather trivial experiments are often quoted because of their simplicity. The reader should not infer that large-scale, expensive experiments are always free of such improper procedure.

67
Watch Out for 67%

"67% of doctors surveyed recommended remedy X" is a statement to be regarded with some suspicion. If the makers of remedy X questioned three doctors and terminated the survey because two of them gave the desired answer, 67% (or perhaps 66%) is what would be quoted. Some people do not look with favor on the expenditure of additional money to get a more reliable answer when they already have the answer they like. If we aren't told how large the survey was, we can't be blamed for suspecting from the result "67%" that only three doctors were interviewed.

Jumping to conclusions is not a new event in the great track meet of life, and it certainly isn't restricted to the advertising brotherhood. The venerability of the line "one swallow maketh not a summer" attests to the fact that making inferences from insufficient data has a long history.

If the makers of remedy X are honest and truthful, how many doctors should they poll before announcing their findings? This question sounds simple but it doesn't have a simple answer. No matter how many doctors we may ask, the inclusion of a few more will make the survey more precise in its attempt to estimate what the result would be if one could reach the entire medical profession. Also, as sample size is increased the additional precision obtained by questioning more doctors becomes smaller and smaller, so that benefits resulting from the additional expense of questioning follow a law of diminishing returns. This and other features of the problem explain why statistical consultants become irritated when clients walk in the door asking, "How large a sample do I need?" before even introducing themselves and taking a seat.

This hypothetical survey of doctors is the simplest kind because the question has only two answers: yes, remedy X is one of the many things I might be willing to prescribe; or no, I wouldn't prescribe remedy X for a burglar. All that is sought is the percentage of doctors who answer "yes." A statistician can determine how large a survey must be so that it will almost surely produce an answer within a certain percentage of the true answer, but the client has to decide what precision is needed. The same considerations come up as in the case of political polls, discussed earlier. If the question to be asked involves quantitative answers, then even more discussion is needed before deciding on a proper sample size. For example, if we want to say, "Doctors surveyed say that high blood pressure should be diagnosed when the diastolic pressure exceeds 90," how many doctors must be asked so that the average of their answers is almost sure to be within 5 percentage points of the average that would be obtained if the entire population of doctors were asked? (And what do we mean, quantitatively, by "almost sure" — and is 5 percentage points close enough?)

Questions such as these can be answered on a statistical basis, but the statistician will not try to answer them until certain information is forthcoming. Before saying how many swallows must be seen before one can safely conclude that summer is here, the statistician will need to learn something about the variability of the migrating habits of individual swallows.

Repeated observations are called "replications" by statisticians. When the results of an experiment are reported in the news, it usually isn't feasible to go into all the details of how the appropriate amount of replication was determined. But readers are entitled to have (and seldom get) some assurance that there was enough replication to produce a meaningful answer.

68
What Experimenters Can Learn from Football and Card Players

Why is a football game preceded by a coin toss? The answer is that this is a "fair" way to decide who gets to receive and who gets to go with the direction of the wind. Fairness is at least as important in an experiment as it is in a football game. If we were comparing the effectiveness of two seasickness remedies it certainly wouldn't be "fair" to try one remedy only on the crew and the other only on the passengers.

One way that fairness can be achieved in experiments is through a process of deliberate randomization, which is simply football's coin toss, sometimes made more complex when the nature of the experiment calls for complexity. If there are 400 passengers on a ship and we have enough seasickness pills to try each kind on 20 people, those people should be selected "at random." To be precise, a random sample of 20 passengers is one selected by a process whereby any group of 20 passengers had the same chance of being selected as any other group of 20 passengers.

Putting the names of 400 passengers into a hat and choosing 40 of them at random takes a little time, and then one has to run down these 40 people and get them to participate. The person carrying out the experiment may decide that it would be much easier to stand on deck and hand out pills to the first 40 people who pass by and are willing to take them. It would probably also ruin the experiment, since the people on deck are not typical—they probably tend to be healthier and stronger than those who are engaged in drinking, playing cards, or being sick in their cabins before the trip even starts. We would have no idea how much these factors would influence the outcome of the experiment. Experimenters who want to avoid the extra trouble sometimes argue that a random selection has a chance of selecting the same subjects that any other method would select, but with random selections we can evaluate the probabilities of drawing wrong conclusions, and keeping these probabilities low is part of the science of good experimentation.

Randomization for the purpose of fairness has probably been a part of games throughout their history, but it didn't enter the field of scientific experimentation until this century when it was introduced and justified by the English statistician Sir Ronald Fisher. To this day experimenters who routinely shuffle and cut cards when playing bridge will often omit the similar precautions that could enable them to reduce the biases in their experimentation.

Randomization is usually a cheap and harmless way of improving the effectiveness of experimentation with very little extra effort. Its dangers are few, although I have heard of one instance in which some experimenters using randomization were accused by straitlaced neighbors of illicit gambling activity!

As usual, the simple example quoted here doesn't give much idea of the various more complex ways in which randomization can be used in experimentation. Most uses of randomization are for reducing biases, but there is one relatively new application that is clever enough to deserve mention. Let's suppose we want to find the percentage of people who have done a particular thing that they don't want to admit to. If we go out and ask men, "Have you ever beaten your wife?," we're not going to get an unbiased result. Suppose, though, that we follow up this question by instructing each respondent to toss two coins secretly before answering. If both come up heads he is to lie, otherwise he is to tell the truth. Since he alone knows the result of the tosses, his answer will not incriminate him individually, but we know that the two coins will both come up heads one-fourth of the time and after asking a lot of men we can get a good estimate of the overall truth. If Y is the fraction of men who give us a "yes" answer to the question, then $2Y - \frac{1}{2}$ is a good estimate of the fraction of wife-beaters in the population sampled. (Incidentally, because of a mathematical quirk, the random device used must not produce lies exactly half the time, and that's why we can't use just a single coin toss.)

69

A Digression on Random Digits
and Computer Simulation

A restaurateur wants to choose a house Chablis from among twenty producers, so he finds ten people representative of the clientele and will ask each of them to taste each of the twenty brands and rank them in order of preference. In what order should the twenty brands be presented to each taster? The order should not be the same for each taster, because there might be some consistent advantage for the brand that was always presented first. Or, the first position could be a disadvantage since the taster, after tasting twenty wines, might forget what the earliest ones were like. Since the restaurateur doesn't know which positions in the sequence are best or worst, and since he doesn't have enough tasters to create a balance, he should opt for random ordering.

This simple example is chosen to illustrate one of the very many kinds of random selections or orderings that may be required in experimentation. What we need here, after we assign a number to each brand, is ten random orderings of the numbers from 1 to 20, one ordering for each taster. It's easy, you might say, to write down the numbers from 1 to 20 in haphazard order and then do this again and again. Human beings, however, are not very good at producing randomness. For example, they tend to avoid ever starting out with the number 1, even though one ordering out of 20 should do so; they also tend to avoid putting adjacent numbers together, achieving too much mixing in an effort to avoid producing too lit-

tle. Asked to produce randomly chosen one-digit numbers, people tend to overproduce 3 and 7, apparently regarding other digits as less random; and when asked to produce two-digit random numbers, they avoid "round" numbers like 10 and 20 and come up too often with 37 and 73.

Before computers became so prevalent, most randomization was done through use of "tables of random digits." Such a table looks like this:

43	23	42	42	17	75	18	33	68	41	72	28	03	57	53
98	75	30	18	01	06	53	76	32	48	61	43	29	86	90 . . .

continuing on for many pages. How these tables were obtained in the first place is too long a story for our purposes. An interesting part of that story is the philosophical argument presented by those who felt that once a set of digits had been written down, it was not random. The important thing is that the numbers "behave" like random numbers (e.g., 5 appears not much more or less often than 8, the digit 3 is followed by the digit 0 about 10% of the time, etc.) After a table passes a battery of such tests, it can be safely

151

used in various ways. To simulate coin tosses, for example, all we have to do is pick a starting place in the table by closing our eyes and putting a finger down "randomly," and then move successively along counting odd digits as "heads" and even digits as "tails."

A crude way of producing a random ordering of numbers from 1 to 20 from a table of random digits is to move through the table and simply reject all numbers greater than 20, as well as any number that has already been met. Thus, if we use the digits in the sample table above in this way, our random ordering starts out 17, 18, 3, 1, 6, ... This process is slow but can be accelerated by taking from each two-digit number the remainder after dividing by 20. With this method the original sequence 43, 23, 42, 42, 17, 75, . . . becomes 3, 3, 2, 2, 17, 15, ... Continuing to do this until all twenty numbers show up, we then strike out the repetitions and have what we wanted.

In large-scale experiments this kind of work can become tedious, so people set out to teach (program) computers to produce random numbers. This effort was successful, although it seemed at first that nothing was less random than a computer. Once it was learned how to generate random digits inside a computer, it became possible to simulate on the computer an enormous variety of real-life activities. The advantage of computer simulation is that it enables us to find out how to improve operations without having to do costly experimentation in actual situations. A bank can simulate customer arrivals at three teller windows and find out how long the lines will be; then it can easily find what improvement would result if four tellers were used. An architect may believe that three elevators are enough for a proposed building, but a computer simulation can predict how good the service will be for a given level of passenger traffic; owners of the building may not consider that quality of service good enough and ask for four elevators, a very difficult improvement to achieve once the building is built.

When a real situation involves chance we have to use probability mathematics to understand it quantitatively. Direct mathematical solutions sometimes exist (as when in bridge we want to know whether it is better to finesse or to lead the ace and king when there are five trumps out to the queen), but most real systems are too complicated for direct solutions. In these cases the computer, once taught to generate random numbers, can use simulation to get useful answers to otherwise impossible problems.

70
Generalizations

"That's a nice looking blue car."
"Well it's blue on this side, at least."

Whether this conversation ever took place or not, it illustrates the fact that some people are afraid to make generalizations. Healthy skepticism can be a virtue, but too much is not healthy. Skeptics like to quote the old paradox, "All generalizations are false, including this one." Yet if we don't generalize, we never learn from experience. Does every object in the world have to be dropped from a height before we believe that all objects fall? What do we conclude when someone drops a gas-filled balloon and it goes up instead of down?

There are two possible actions after observing the rising balloon. One action is to reject everything that has been learned before and to claim that we really don't know whether an object will fall or not, since some do and some don't. (Perhaps no one ever took this position on this particular question, but it has certainly been taken on more difficult questions.) The other action, naturally, is to find out why the balloon rose, and then to conclude that objects heavier than air fall, while objects lighter than air rise. The invention of the airplane causes us to go through still another exercise of this sort. We may never come to the end of these adjustments to our understanding, but at no time do they contradict the fact that if we release a block of wood it will fall. (Unless, of course, we happen to be standing underwater at the time.)

Very often the generalization we want to make consists of making a statement comparing one group of individuals (not necessarily people) with another group, when the individuals within the group vary widely. For example, the Rhode Island Red hen is said to be a good egg producer, while the Dark Cornish hen is not. Statements like this are an important part of our knowledge and experience, yet their truth is often challenged by legalistic types. Some Rhode Island Reds are better layers than others, and similar variation occurs among Dark Cornish hens, so it is probably possible to find among the most prolific Dark Cornish hens at least one that lays better than at least one Rhode Island Red. Therefore, if we take the statement "Rhode Island Reds are better layers than the Dark Cornish" to mean that every Red is better than every Cornish, then this generalization is false. But that is not what this generalization means: what it means is that the *average* production of Rhode Island Reds is greater than that of the Dark Cornish.

In cases where averages aren't meaningful it isn't easy to explain what a particular generalization means, but generalizations are nevertheless possible and useful. For example I can say, "Flowers are prettier than weeds," and if we can agree on what "pretty" means we may be able to agree on this statement. We can't average prettiness, and we certainly can find some weeds that are prettier than some flowers, but we can agree on the generalization if it means that most flowers are prettier than most weeds.

Bitter arguments often start because one person has made a generalization and a listener chooses to interpret it as claiming to be universally true in every individual instance. Once this point has been explained the argument often disappears. There are cases, however, in which the comparison between two groups is fairly close while the variation within each group is comparatively large. When this is true, it can be argued that an observed difference in averages may be merely a sampling accident, and that another sample of observations could reverse the conclusion. At this point statistics comes to the rescue, as we shall subsequently see.

71
Within vs. Between
From Toothpaste to Static

For the moment let's be an executive of a company that makes toothpaste. Our chemists and biologists have decided from their research that an additive called "Q" should be helpful in reducing tooth decay. We decide to test this idea on the teeth of living human beings. How should we go about it? We could try Q on our own family for six months, observing perhaps one cavity among the four of us. If in the previous six months we had two cavities, is the reduction meaningful? It is not, because there has often been more variation than this in the number of our cavities even when no change was made in our toothpaste. What we need is more people, to "average out" the chance fluctuations.

We then propose to go out into the world and give Q toothpaste to hundreds of people, intending after twelve months to see how many cavities they developed. But this result must be compared with something, and most subjects would probably be unable to tell how many cavities they had had during the preceding twelve-month period. We need some "controls." We will enlist 1000 people to use Q toothpaste and another 1000 to use whatever they have been using. How do we know that 1000 is enough replication? As pointed out before, "That's all we can afford" is not an acceptable answer to this question. Neither is, "1000 is such a big number that surely everything must even out." The question should be addressed to a statistician, who may raise some other questions before giving an answer.

If the statistician's questions can be answered and it turns out that 1000 people in each group is enough, there are still precautions that must be taken. A psychologist tells us that subjects getting the Q additive may respond enthusiastically by being particularly careful and energetic in their brushing; if their cavities are fewer than those of the controls, it could be the effect of the extra brushing. We realize that a "blind" experiment is called for, so we decide to set it up in such a way that the subjects don't know whether they have Q in their toothpaste or not.

In planning the experiment, we should anticipate how the results will be analyzed. When the results are in we will undoubtedly notice a great amount of variation. The first thing we will look at will be the difference between the average number of cavities of the Q users and the corresponding average for the non-Q users. Statisticians call this the "between" variation. If the Q users have fewer cavities then, as company executives, we may want to rush into print. Statisticians, however, will be more cautious. They will point out that there is a lot of variation within each group, and they call this the "within" variation. Since this variation exists, the result of the experiment might be an accident, and a confirmatory experiment tried on another 2000 people might produce the opposite result. So mathematical techniques will be needed to determine whether a "between" variation as large as that observed could have a reasonable chance of happening because of the amount of "within" variation that was present. If the answer is "no," then the result will be declared "statistically significant" and the additive Q can be advertised as being effective with very little risk that later observations will embarrass us.

In general when the between variation is large compared to the within variation we don't need a large number of subjects to be convinced of the validity of a result. That is, if the number of cavities doesn't vary much from person to person, and if the effect of Q is great, we can see a convincing result without using thousands of people. But if the variation from person to person is large, and the effect of Q is not very great, it takes a very large number of subjects to make the within variation "average out" to the point where the small effect of Q can be correctly detected.

Interpreting data under such circumstances is like listening to a radio during a thunderstorm. The between variation corresponds to the information we want to hear from the radio station. The within variation corresponds to the unwanted static noise that interferes with the station's signal. When the static is loud we need a stronger signal from the station in order to understand correctly what is being said. If the station doesn't have a strong enough signal, it can achieve the same effect by being repetitious, or redundant, as exemplified by repeating the name of a sponsor. This repetition performs much the same function as replications in experiments. People in communications engineering speak of the "signal-to-noise ratio" to describe this situation. If the signal-to-noise ratio is very low, you can still get the meaning of the signal if it is repeated often enough.

Experiments usually are looking for "signals" of truth, and the search is always hampered by "noise" of one kind or another. In judging someone else's experimental results it's important to find out whether they represent a true signal or whether they are just so much noise.

72
A Thought About Stereotypes

As a southerner who has spent much of his life in the north I know that some northerners tend to think of southerners as unmotivated, ignorant, slipshod, or lazy. And some southerners tend to think of northerners as rude, boorish, overaggressive, and generally unpleasant. Northerners like to think of themselves as efficient and businesslike, while southerners like to see themselves as polite and gracious. These extreme images become much less intense when people migrate and intermix and get to know more individuals of the opposite group. Experience with people always shows that there is great variation within any group even if the group appears to be homogeneous to outsiders. (Statisticians commonly shorten the phrase "variation within a group" to "within variation.")

After we have learned that there is within variation (some geniuses are much smarter than others) it is a mistake to conclude that variation between groups doesn't exist. Experience in experimental science teaches that whether we're dealing with people, animals, plants, metal bars, or batches of chemical, if two groups are identifiable as different there will be differences between them in every measurable characteristic. If we could measure efficiency and graciousness, we might indeed find that northerners are on the average a little more efficient than southerners and that southerners are on the average a little more gracious than northerners. These "between differences" (differences between the group averages) may be small compared to some of the "within differences," but if they were not there at all one might wonder where the images originated, and why it is that they seem to prevail with minor modifications in other countries. (It could be strictly a reaction to climate, and I've often intended to ask my Australian friends if the phenomenon is seen in reverse in that global control group of the other hemisphere.)

The within-vs.-between question pops up frequently in everyday life. There we often can't apply the numerical methods of statistics to it, but many discussions can be clarified by use of the terms "within variation" and "between variation."

73
Longevity at the Hot Corner

Did you know that major league third basemen have a mortality rate that is 12% lower than that of their teammates and 45% lower than that of the general male population? This mind-boggling fact surfaced in statistics gathered by the Metropolitan Life Insurance Co., according to Stan Isle's column in *The Sporting News,* November 15, 1980. Although you may think this falls short of being the most fascinating fact you've learned this month, nevertheless it makes a good subject for study, since the statement is very similar to ones of more serious import that we read about daily.

First of all, this is a good example of the within-vs.-between situation. People die at widely varying ages, and third basemen are surely no exception, so the within variation is high. The between variation is the difference in true mortality rates from one position on a team to another. Assuming that there is no real difference in the mortality rates of players at different positions, it is clear that if we look at a certain group of players (those who have been in the major leagues during the last twenty-five years, or those who hold policies with a given life insurance company, or whatever) we will observe, by chance, and because of the within variation, different mortality rates for each position. The question is, would we or would we not in these circumstances observe differences as large as the 12% reported for third basemen in the study? Only Metropolitan knows for sure, since neither *The Sporting News* nor any other publication is likely

to delve into the statistics deeply enough to find out. Yet if we don't address ourselves to this question, why report the result? It is of no interest to know that the particular third basemen covered in the study have an unusually low mortality rate unless we have evidence to show that the next study on the next group of players will also come out more or less the same way.

Now suppose we have examined the statistics carefully and have concluded that in truth third basemen do live longer than their teammates. Should parents advise their sons and daughters to try out for the hot corner? ("Hot corner" is an ancient baseball cliché for the position, alluding to the fact that third basemen have to face more hard-hit balls than others do.) More to the point, should parents advise their children to take up baseball and get in on the 45% lower mortality rate? They would if they follow the same reasoning that sends children to college merely because college graduates on the average make more money than others. All the benefits of regular exercise aside, it seems obvious that good health and consequent longevity help to make an individual a major league baseball player much more than vice versa.

In addition to the within-vs.-between aspect of this story, mortality statistics are tricky and conclusions drawn from them should always be viewed with skepticism. People have been known to look at our marriage and divorce totals for a year, note that the latter is about half the former, and conclude that half of all marriages will end in divorce. It also happens that we have about half as many deaths as births each year, but we don't conclude from this that half of all the people born will die.

74
Let's Save Some Money

Research costs money and it is paid for by all of us, as taxpayers and consumers. We probably get more benefits from it than from many other things that are bought for us, but we still want to see that money spent on research is not wasted. So while we are talking about protecting ourselves against incorrect conclusions from data, it is appropriate to say a few words about how our money is spent to get the data.

Unfortunately, research is not like a trip in which we know our exact destination and have a map that shows us the very best way of getting there. It's more like Columbus's trip: we don't necessarily know what our destination is, we ¬ay or may not find it, we may find an even better destination, or we may fall off the edge of the earth. Nevertheless, there are efficient and inefficient ways of proceeding in research, and more attention needs to be given to looking for the efficient routes. Statistics has provided a number of very helpful methods for improving efficiency, one of which, called "blocking," is both important and simple.

The basic idea of blocking is that experimental material (people, animals, plants, plots of ground, chemicals, whatever) can usually be separated into blocks within which there is less variability than there is over the whole material. When this can be done, experimental comparisons can often be made much more efficiently within these blocks than if the existence of the blocks is ignored. The hypothetical toothpaste experiment mentioned earlier provides a simple example of the general ideal of blocking:

In the test to determine whether the Q additive helped reduce tooth decay, the great variability in the number of cavities from one person to another made it necessary to include large numbers of people in the test. If only people were more alike we could get an equally reliable result without the expense of involving so many of them. This raises the question, "What people are the most alike?," to which the obvious answer is, "Identical twins."

Identical twins have the same heredity and usually similar diets and oral bacteria — all important factors in determining rates of tooth decay. If we have one of a pair brush with the Q additive and the other without, these three sources of variability will not obscure the effect of Q. There will still be some variation, of course, from one of a pair to the other, so several sets of twins must be used to produce conclusive results. Even though finding identical twins adds a little cost, the resulting reduction in the overall size of the experiment saves considerable expense.

In some areas of science experimenters have made extensive use of the idea of blocking and so have saved money or improved the reliability of their results. In other areas people have been more backward and have failed to take advantage of this idea. Taxpayers and consumers have a stake in seeing that this relatively simple procedure becomes more generally used.

75
There Are No Easy Solutions

If you believe everything you are told by the media, your Gullibility Quotient is 100. If you don't believe anything they tell you, your GQ is 0 but you probably have a closed mind and rarely learn anything new. Minimifidianism (having little or no faith) can be as unproductive as gullibility. In steering a course between these two extremes, readers of this book may notice that the problem can be described in terms of Type I and Type II errors:

Believing something that is not true is a Type I error. Not believing something that is true is a Type II error. Bombarded as we are by news, facts, stories, myths, fiction, and outright lies, how can we sensibly decide what to believe?

Different categories of news seem to contain different percentages of truth. If we believe 30% of what is said about the private lives of film and television stars we're probably giving the media the benefit of the doubt. At the other extreme, media reporting of science is considerably more reliable. This, however, doesn't mean that we should swallow everything that pertains to science, or everything that has the endorsement of someone who is or appears to be a scientist.

A good piece of general advice is to remember that THERE ARE NO EASY SOLUTIONS. This may not be quite true, but it ranks pretty high on the truth ladder compared to most pithy sayings. What it says is that nearly all of the easy solutions have been found, and not just recently. It says to beware of those who are selling an easy solution to an old problem, especially if the problem involves people. It says, in particular, to beware of people who are using statistics to push their easy solution to some problem.

Believers in easy solutions tend to polarize into warring camps. If a certain pill is tested and no beneficial effects are observed, one camp always wants the federal government to crack down on the manufacturer while the *caveat emptor* camp always wants the government to stay out of it. Each camp has a single easy solution to all problems. My hope is that readers of this book will think of the trade-offs involved — whether the government steps in or not should depend on what the pill is supposed to do, how harmful its side effects may be, what testing the pill costs, how reliable the tests are, and whether the government does or does not have better ways of spending its money.

Testing is often the only way to answer our questions, but it doesn't produce unassailable, universal truths that should be carved on stone tablets. Instead, testing produces statistics, which must be interpreted, and usually the less willing we are to risk false interpretations the more money we must spend on the testing. Good statistical practice makes testing more efficient and so less costly, but hard decisions about trade-offs are always with us.

Not so long ago it was not uncommon to hear even a scientist say, "I don't believe in statistics." This was in part a natural reaction to the many incorrect conclusions that people draw from statistics, but it was also an "easy solution" to the scientist's problem of rationalizing a reluctance to learn about new developments in the field of planning and analyzing experiments. Today most scientists realize that they *are* using statistics — their only choice is whether to use good statistical methods or bad ones.

76
Separating the Liars
from the Statisticians

Hardly a week goes by that doesn't produce statistics that would inspire a few more pages for this book. The book was intended to be short and simple, however, and its point has been made, so this seems to be the time to stop.

Much more has been said on all of these subjects by authors who write longer books for students and practitioners. This book never claimed that it would make you a statistician. It also never claimed that telling the liars from the statisticians is easy. It crossed my mind to try to make it easier by concluding with a checklist — a summary of things to look out for if you are a consumer of statistics. However, the rich variety of tools available to the data pusher would make the list so long that few would read it. The reader who wants a summary can get one by scanning the table of contents.

My goal has been reached if I've shown that, even though you may still find numbers themselves dull, questioning numerical results and interpreting them can be a fascinating and rewarding activity.

Index

Index

Index

Index